Hormones and Reality

John S. Torday

Hormones and Reality

Epigenetic Regulation of the Endocrine System

Springer

John S. Torday
David Geffen School of Medicine
University of California, Los Angeles
Los Angeles, CA, USA

ISBN 978-3-030-93690-7 ISBN 978-3-030-93691-4 (eBook)
https://doi.org/10.1007/978-3-030-93691-4

This Springer imprint is published by the registered company Springer Nature Switzerland AG
The registered company address is: Gewerbestrasse 11, 6330 Cham, Switzerland

Preface

This book is founded on the breakthrough concept that the cell-cell signaling mechanism for embryologic development can be traced backwards across space and time from present-day physiologic traits to their origins, both biologic and physical. In the larger context of cell-cell signaling as the basis for both embryologic development and phylogenetic speciation, there must be a modicum of truth to this supposition. Particularly when the thesis of this book – that the endocrine system connects us to the Cosmos – is both realizable and realized by hypothesis-testing scientific experiments. Suffice to say that superimposing the cell-cell signaling of phylogeny on development shifts the frame for the process of evolution from "endless forms most beautiful" to the flow of energy, moving away from Darwin and towards Whitehead Process Philosophy as a paradigm shift.

Over 100 peer-reviewed scientific journal articles and 5 monographs have been published based on this concept to date.

Unlike Darwinian evolution, which is predicated on random phenotypic variation as the source of biologic adaptation, the variation in cell-cell signaling is caused by the loss of homeostatic control, leading to cellular remodeling, culminating in the re-establishment of homeostasis with reference to the First Principles of Physiology. The aforementioned is the result of life being self-referential and self-organizing. The capacity to restore cellular homeostasis is contingent on serial exaptations or pre-adaptations – the repurposing of genes previously used successfully in other existential contexts over the course of the evolution of the organism. As such, the "history" of the organism can be re-enacted, revealing how and why changes in physiologic traits have occurred in response to environmental stresses. Why the "greenhouse effect" caused by the accumulation of carbon dioxide in the atmosphere resulted in environmental warming, causing the partial drying up of the waters covering the Earth, exposing land masses, and depleting those waters of oxygen. As a consequence, our boney fish ancestors were forced out of water onto land, doing so on at least five separate occasions according to the fossil record, as a "trial-and-error" "experiment of nature." It is only because of the cell-cell signaling basis for our physiology that such empiricism was possible, "inscribing" us with self-knowledge step by step. Indeed, life is our handbook, and the history of that on-going effort is written in our physiologic evolution. It is that story that is being related in this book, informing us why we feel there is something greater than ourselves.

Given enough data linking environmental change with genetic retooling, the cell-cell signaling model for physiology leads to predictive biology, finally rendering biology a "hard" science, on par with physics and chemistry. This book tells that tale in its many iterations. The hope is that the messages conveyed in its chapters will resonate with the reader, so that we may move forward together in a new narrative that departs from the traditional false narratives we have been telling ourselves to cope with the ambiguity of our origin in negentropy, as first related by Schrodinger in *What is Life?* That revelation resolves the conundrum set forth by Heraclitus, for example, telling us that we cannot step into the same river twice, yet our attempts to do so have now paid off by using the Scientific Method of duplication and replication. It also explains the importance of Heisenberg's Uncertainty Principle, providing a scientific premise for our ambiguous origin – "like dissolves like." In that vein, the poet Robert Frost stated in one of his notebooks that "Life is that which can mix oil and water."

The idea of resolving our ambiguous existence echoes Carl Jung's concept of Synchronicity as a manifestation of the "collective unconscious." Or John Lennon saying "A dream you dream alone is only a dream. A dream you dream together is reality" – that sense of transcendence expresses what David Bohm describes as the Explicate Order, formed by our subjective senses, always reaching for the Implicate Order through experimentation. That is life, dear reader.

The **Secret Sits, by Robert Frost**

We dance round in a ring and suppose, But the Secret sits in the middle and knows.

from Little Gidding, by T.S. Eliot

... We shall not cease from exploration And the end of all our exploring Will be to arrive where we started And know the place for the first time.

Philadelphia, Pennsylvania, 2021

Los Angeles, CA, USA John S. Torday

Contents

Chapter 1
The Phenomenon of "Subjective Age" as an Epigenetic Cellular-Molecular Mechanism

Introduction

There is empiric evidence that the sensation of a differential between a person's subjective judgment of physiologic age and his or her actual chronological age is a nearly universal experience that shifts over the life cycle. During adolescence, individuals begin feeling older than their chronological age (Montepare and Lachman 1989), whereas in early to mid-adulthood the phenomenon of subjective age reverses, individuals feeling younger than their chronological age (Shinan-Altman and Werner 2019). It has been calculated that from midlife on, individuals feel about 20% younger than their actual age (Rubin and Berntsen 2006). By inference, paradoxically, the older you get, the younger you feel.

In a series of prior articles and books, a framework of biology and evolutionary development has been presented that concentrates on the importance of cell-cell communication among self-referential cells (Torday and Rehan 2012). It is now argued that this framework can better explain the phenomenon of subjective age as a product of the self-referential cellular assessment of current homeostatic equipoise among individual cells referenced to cellular standards that originate within unicellular origins as aggregated through cell-cell communication to the level of the total organism.

Unicellular Origins, First Principles, and Ambiguous Information

When lipids were transported to earth by snowball-like asteroids, they were immersed in the primordial ocean that covered the earth and spontaneously formed micelles or cell-like spherules (Groen et al. 2012). External and internal

J. S. Torday, *Hormones and Reality*, https://doi.org/10.1007/978-3-030-93691-4_1

environments were delineated by being surrounded by a semipermeable membrane that selectively allowed certain molecules in and out of the cell (Schrum et al. 2010). This specific combination of cellular features forms the basis for evolution perpetually referencing the first principles of physiology (negentropy, chemiosmosis, homeostasis) (Torday and Rehan 2009), which permit the faithful endogenization of the external environment. This is the hallmark process for evolutionary development, as it enables endosymbiosis, which initiated eukaryota, and forms the basis for self-referential self-organization as a perpetual reciprocal with the singularity (Torday 2019). It has been previously advanced that the homeostatic force that instantiates self-referential self-organization is the "equal and opposite reaction" to the Big Bang that produced all matter in the cosmos (Hawking 2011). Simultaneously, it generated the ground state conditions by which homeostatic forces ultimately yield biologic materialism. In so doing, a set of primordial first principles of physiology was established for the living state, providing a set of initiating essential cellular reference points that perpetually guide the living circumstance.

As one of the triadic cornerstones of the first principles of physiology, homeostasis is the linchpin of self-referential self-organization. This consistent attempt to maintain homeostatic equipoise defines cellular life (Torday 2015b). Nonetheless, the living state is obviously dependent on the appraisal of information and its communication. However, sources of information that are available to living systems are always imprecise. The reasons are threefold. Any resources that a cell might deploy on the basis of information must be a contingent self-referential choice among a range of possibilities (implicates) from which explicate cellular actions might spring (Miller Jr et al. 2019). Secondly, since living entities are self-referential observer/participants, by definition, information in the self-referential state is never exactly alike to any two observer/participants or to any single observer making repeated measurements of the same phenomenon (Miller Jr et al. 2020). In this regard, living information is better conceived as a volume than as a point, the sides of which are open to variable interpretation dependent on observer position with respect to any environmental cue. Thirdly, all information that any cell can receive for appraisal or communicates to others must be transmitted across boundaries via intervening media with inevitable time delays. Thus, all cellular information and communication are subject to unavoidable degradation. As a result, living systems exist within an innate state of ambiguity (Torday and Miller Jr 2017). It follows that cellular life and evolutionary development are self-organizing cellular responses to that definitional uncertainty. As a critical derivative, cells must continuously conform to a set of basal initiating parameters based within the first principles of physiology, which themselves extrapolate from the singularity, or they would inevitably drift over time due to successive accumulation of ambiguous environmental cues. The cellular answer to this problem is twofold. As a measuring instrument, each cell must adhere to the basic first principles of physiology, which underscores its intrinsic self-referential measuring apparatus. Secondly, to further prevent possibly fatal skewing from those principles by the accumulation of epigenetic impacts, all multicellular eukaryotic organisms undergo an obligatory recapitulation through a unicellular zygotic phase in which those garnered epigenetic impacts are adjudicated during the reproductive process (Maamar et al. 2021).

First Principles of Physiology

One source of failure to advance biomedicine has been the unrewarded expectation that the Human Genome Project (HGP) would reconstruct physiology from genes; but physiology is not due to genes communicating with genes; physiology is the product of cells communicating with cells (Demayo et al. 2002). Just as the atom is generally considered the smallest functional unit of material physics, the cell is properly considered the smallest functional unit of biology. Trying to comprehend gene regulatory networks based on individual genes without regarding that cohesive functional context results in elements of biologic action that do not represent physiology. What is required then is a better means of translating genes and their identifiable properties into physiologic principles, like the use of the periodic table to "translate" the physical properties of the elements into chemistry (Scerri 2019).

For example, the evolution of the lung can be "deconvoluted" by applying cell-cell communication mechanisms to all aspects of lung biology – development, homeostasis, and regeneration-repair (Torday and Rehan 2007). In this frame, gene regulatory networks that are common to all of these processes can be better used to predict ontogeny, phylogeny, and the disease-related consequences of failed cell-cell signaling. This algorithmic approach elucidates characteristics of vertebrate physiology as a cascade of emergent and contingent cellular adaptational responses, rather than as random genetic mutations (Darwin 1859). It is maintained that by mapping complex physiologic traits onto gene regulatory networks, and arranging them akin to the periodic table of elements in physics, the first principles of physiology, upon which all cells depend, will emerge.

Physics and Physiology

David Bohm hypothesized that there are both an explicate order and an implicate order in the cosmos, the former being our subjective view of the latter due to our evolved senses (Bohm 1980). Both of those states of being are present within the organism, the explicate acting as the drive for seeking epigenetic "marks" in the environment that constitute changes that pose a threat (Torday and Miller 2016). The endogenization of such marks has formed our physiology. In turn, all explicates must first arise from the superimposition of possibilities contained within the implicate order. The interplay between these two orders provides the means by which we are able to evolve in concert with our ever-changing environment based on the first principles of physiology, which are themselves constrained by the initiating conditions of the singularity.

From this background, a cellular dynamic accounting for subjective age can be identified. Cells exist in circumstances in which the information upon which they depend is imprecise. The assessment of that information and the choice of whether or not it will be communicated is necessarily a measuring process. Further, that measuring process must be judged according to some standard from which a measurement can be made, which for cells is their conformity with first principles (Torday and Rehan 2009). Hence, all cells are consistently appraising themselves and their state of homeostatic equipoise. At all times, cells are weighing implicates and explicates to be prepared for further action and, importantly too, communicating that status to their cellular ecologic companions (Torday 2016). This process of cell-cell communication forms the basis for the phenomenon of subjective age based on the cellular accumulation of environmental experiences.

Mechanisms of Development as a Continuous Cellular Interface with the Environment

Development from the zygotic fertilized egg stage forward is mediated by soluble growth factors as signals for cell-cell interactions between cells of different germ line origins (endoderm, ectoderm, and mesoderm) (Torday and Rehan 2012). The aggregate of these sensory interactions with the external environment has been expressed as the senome (Baluska and Miller Jr 2018), which is the integration of the totality of the sensory information inputs to cells to generate form and function. As the zygote morphs from the blastula to the morula and gastrula, Wolpert has said that "gastrulation is truly the most important time in your life" (Wolpert and Vicente 2015). That is because it is the stage at which the mesoderm is introduced between the endoderm and ectoderm during embryogenesis (Wolpert 1992). The mesoderm adds plasticity to the developing conceptus under the influence of both physical and chemical factors that introduce change in response to the environment (West-Eberhard 2003). That is particularly true of the endocrine system affecting the conceptus, since the endocrine system is also under epigenetic control.

It is further advanced that the APGAR score is a practical cellular measure of this dynamic interface. The APGAR score is a systematic means of evaluating the physiologic development of the newborn (Apgar 1953). As such, that "score" would also reflect the cellular integrity of the newborn that aggregates as its consciousness and integrated physiology. For example, preterm infants cannot effectively maintain their body temperature, and the evolution of body temperature control is a hallmark of vertebrate evolution, including consciousness (Torday 2015). Bergson defined consciousness as "thinking of the past and planning for the future" (Jancsary 2019). The newborn lives in the present, since it has no past and cannot conceive of the future, so it does not experience the environment on a scale of "subjective age." An infant behaves like Aristotle's "blank slate," maximally absorbing epigenetic data from its environment. For instance, Piaget stipulated that the infant had to experience specific stages of development in order to accommodate our large brains (Piaget 1977). It is maintained that this stepwise interrelationship with the environment, proceeding from the mother's breast, to crawling, and then to fuller ambulation, is an efficient means of allowing the infant a developmental period to assimilate its initial epigenetic experiences as its own process of environmental endogenization.

Although phenotypes are conventionally assumed to represent biologic traits or physical features, they should be properly interpreted as networked agents of an organism to obtain epigenetic "marks" from the environment that affect adaptation (Torday and Miller 2016). It is now acknowledged that epigenetic inheritance is a major mechanism by which the environment interacts with the genome. This epigenome is further mediated by germ line cells during meiosis and the subsequent stages of embryologic development (Maamar et al. 2021). Consequently, the phenotype is the biologic manifestation of the active construction of cellular ecologic niches for the active acquisition of epigenetic marks (Torday 2016). Thus, it is a dominant evolutionary force, not merely a passive consequence of Darwinian selection for reproductive success. Reproduction can then be reconsidered as a dynamic frame in which epigenetic inheritance affects growth and development in continued reciprocation with environmental stresses. The obligate return to the unicellular zygotic form can now be reinterpreted – absent a perpetual re-centering to the first principles of physiology to determine the limits of epigenetic inheritance, cellular life would be fatally skewed by overreactions to merely transient environmental conditions.

One popular theory of human evolution is that we are neotenous primates that maintain an immature state of development, accounting for our disproportionately large heads and relatively hairless bodies (Gould 2002). To accommodate the positive selection for our large heads and their contents, humans are born with an immature brain in order to pass through the birth canal. As a consequence, we are born with a brain that is immature, being only 25% of its mature size at birth.

Pathologically, being born small for gestational age results in precocious adrenarche (Novello and Speiser 2018), the adrenal gland producing so-called weak androgens (dehydroepiandrosterone, androstenedione), which are not masculinizing but nevertheless initiate the process of puberty. Such phenomenology may underly subjective age, given the close interrelationship between sexual development during adolescence in association with feeling older than our chronological age (Montepare and Lachman 1989) and the loss of sex hormones in later life in association with feeling younger than our chronological age; in adolescence, androgens from the testes and adrenal cortex increase (Hiort 2002), whereas in mid- to late-life androgens from both sources wane (Morley 2001), suggesting that androgens stimulate the psyche's sense of maturity as youths, whereas in mid- to late-life the loss of androgens makes us feel younger (Wettstein et al. 2021), perhaps to maintain our zest for life or alternatively, as a mechanism for ensuring continued group acceptance despite diminishing physical vigor and survival advantage.

Functionally, puberty impacts on "risk taking," which would tend to both enhance the collection of epigenetic marks from the environment and serve to attract attention and, if successful, increase the youth's social standing within the group (Collado-Rodriguez et al. 2014). Conversely, during later life the obtaining of epigenetic marks is seemingly superfluous, given that we are beyond the reproductive stage; yet it is important to maintain both an inward and external appearance of "health" as our external appearance shows our chronological age to wit the mechanism of subjective age – beginning in mid-life dehydroepiandrosterone levels decline as we age.

This convoluted mechanism begs the question as to why sex hormones should hypothetically cause subjective age. Yet it should be pointed out that it is the ovaries and testes where epigenetic marks are processed (Maamar et al. 2021), providing a logic for this putative mechanism since the hormonal secretions of the gonads determine the experience of subjective age in adolescence and late life, respectively. The further effect of the adrenal androgens is perhaps more challenging to understand in this context, yet Porges' Polyvagal theory might be instructive (Porges 1995). He has invoked the evolution of the vagus nerve in its integrative effect of the adrenals on the heart and brain as a means of mediating emotion. This role of sex in subjective age is only one of many such influences of sex on physiology, dictating the role of the phenotype as agent (Torday and Miller 2016).

A Holistic Approach to Subjective Age

In order to understand the otherwise counterintuitive phenomenon of subjective age, depicted schematically in the accompanying Fig. 1.1; a "first principles" approach is insightful. The formation of cellular boundaries engenders life through negentropy, supported by chemiosmosis and controlled by homeostasis, termed the first

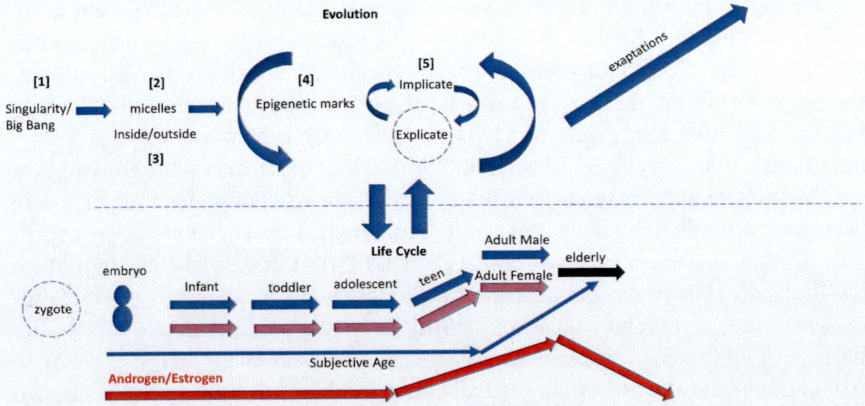

Fig. 1.1 "Exaptation of Subjective Self." (Upper field) The cosmos was formed by the singularity/Big Bang **[1]**; 13.8 billion years ago the earth formed and was pelted by snowball-like asteroids that formed the ocean; lipids present on those asteroids spontaneously formed micelles **[2]** or primitive cells, delineating the inside and outside of those cells **[3]**. The first cells (circle with dotted border) allowed entry of factors in the environment or epigenetic marks **[4]** as the forerunners of physiology, delineating the explicate from the implicate order **[5]**. (Lower field) During development, the zygote recapitulates evolution, and postnatally the infant again acquires epigenetic marks. Hormonal effects (sexual differentiation) perpetuate environmental epigenetics. During adulthood there is a partitioning of external chronological appearance from internal sense of age, referred to as our subjective self

principles of physiology (Torday and Rehan 2009). Awareness of that state as self-referential self-organization arises from homeostasis. The preference for homeostasis depends on the appraisal of information and its communication. However, sources of information and their dissemination are always imprecise. As a result, all living systems exist within an innate state of ambiguity (Torday and Miller 2016). Cellular life and evolutionary development are a self-organizing cellular response to uncertainty, conforming with its basal initiating parameters iteratively (Torday and Miller Jr 2017).

Conventionally, this process is referred to as exaptation, Gould and Vrba's explanation for the repurposing of earlier genetic traits for new applications (Gould and Vrba 1982). In the case of subjective age, it references the path from lipid micelles to cholesterol and molecular memory to the endocrine system, synthesizing sex hormones from cholesterol. The supervening operating principle is the first principles of physiology, which are adhered to through development and phylogeny in order to remain in compliance with the laws of nature.

The true nature of pleiotropy as the distribution of the same gene among different tissues of the body reveals the underlying mechanism of exaptation (Torday 2018). Actually, it is the repurposing of such genes over the arc of the evolution of the organism as exaptations. This process is mediated by cell-cell interactions governed by homeostasis, translating physiologic stress into allostasis (McEwen and Wingfield 2003). The cell, tissue, organ, and organism level interactions are all coordinated by the core first principles of physiology governing mechanism for integrated physics and biology.

Discussion: Subjective Age and the Vertical Integration of Physiology

The foregoing outlines a constant reciprocating dynamic between an inside and an outside, based on first principles. These principles themselves are derived from within the physical parameters imposed by the singularity, which perpetually conditions self-referential self-organization, epitomized by the cell (Torday and Miller Jr 2018). Importantly though, this separation must be maintained within an obligatory context of ambiguous informational cues from the environment (Miller Jr and Torday 2018). Necessarily then, the reception, assessment, and deployment of information must process through self-referential cellular measurements (Miller Jr et al. 2019). In multicellular organisms, this must translate into the essential cell-cell communication that is the active means of the multicellular living state and its further evolution.

It can be argued that inside-outside is merely any simple boundary within living systems. Instead, it should be perceived as analogous to Bohr's principle of complementarity, which highlights particle-wave duality (Bohr and Rosenfeld 1996). Niels Bohr's explanation for that duality is that the phenomenon is a function of the manner in which it is measured. It is proposed that this same situation equally applies to the living state. The cell assesses the external environment and adjusts its internal physiologic milieu in response to it. In turn, it reciprocally affects the external environment. This is an obligatory reciprocation in the self-referential frame. In an observer/participant construct, all explicate biologic actions are information to any other self-referential entity within its information space (Torday and Miller 2016). Any action by a cell is work, and that work provides an informational signature to any other observer/participant within its informational network. In this manner, all cellular actions are in continuous communication with the external milieu, outside of its own membranous boundaries, which receives its actions as information and initiates a further set of reactions among other participants/observers. Thus, cellular boundaries are not so much barriers as drumskins, which beat according to environmental stimuli. This mandated reciprocation yields a process of mutualizing niche constructions that form the essence of the living state (Torday 2016). It is exactly in this manner that epigenetic marks become a process of continuous endogenizations of the external environment, and what is exactly "inside" and what is "outside" depend on how the measurements are construed, i.e., which observer/participant is doing the measuring.

The sensation of a differential between a person's self-assessed perception of age and her/his chronological one as subjective age is a well-documented and nearly universal phenomenon among humans (Alonso Debreczeni and Bailey 2021). It is also well-known that this sensation varies across our life cycle stages. In middle age and the later adult years, individuals report a generally younger subjective age than their chronological one. Inversely, in adolescents and teens, subjective age is judged as greater than chronological reality. In infants, the differences between actual and subjective age is effectively nil. Theorists have contended that aging adults maintain subjective age as a means of defensive denial of the aging process and the stigma which attaches to it. Subconscious denial of aging has also been seen as an adaptive mechanism that defends a psychological adjustment to aging that is presumed to confer health benefits, as well as social benefits (Kwak et al. 2018). Others have attempted to model self-esteem based on a scoring system in which financial satisfaction across middle age is cast as a relevant mediator of differences in the perception of subjective age (Bergland et al. 2014). It is argued here that the problematic issue of subjective age can be illuminated by utilizing a single unifying approach based on cellular dynamics and relevant cell-cell communication in response to aggregate cellular epigenetic experiences.

As noted above, our origins derive from the formation of micelles from lipids in the primordial ocean, separating the internal environment of the cell from the outside environment. Claude Bernard referred to this as the milieu interieur. The basis for this successful separation is the consistent and continuous endogenization of the outward environment by cells that matches a measured adherence to the first principles of physiology. Memory is critical to this process, which itself is believed to have originated in the primordial cellular stage as lipid hysteresis. Thus, the lipid-containing cell membrane serves not only as a barrier, and a reciprocating participant in inside-outside dynamics through active chemiosmosis, but is also serving as a form of memory. All are parts of the process of the cellular measurement of current status compared to fixed reference points of the first principles of physiology. In this manner, the cell, as a measuring apparatus, judges its current state versus a form of "objective" status.

It is pertinent that there are two coexisting and connected clock cycles that have been identified in cells; one controls cell division and the other acts as a circadian pacemaker (Mohawk et al. 2012). In multicellular organisms, both of these link to the various cellular ecologies and physiologic processes that sustain life. Thus, cellular timekeeping and a continuous assessment of status, both within the present moment and through memory, connect to transcriptional and posttranscriptional cellular feedback loops to maintain cellular homeostatic equipoise. It follows that insofar as each individual cell has this imposed self-referential frame, then aggregate multicellular organisms must also. As a general phenomenon, this is manifested through our obvious circadian rhythms. Yet, the same dynamic implies a general cellular-based organismal sense of its collective life cycle, which it experiences as its personal perception of aging.

From this, it is asserted that subjective age is a function of the combination of evolutionary requirements to support the entire organism and the real-time experiences of the cells that constitute it. For infants and the very young, there is no gap between subjective assessment and chronological age, as cells have not accumulated enough environmental experiences to discriminate between potential differential states.

For teens and young adults, there is an evolutionary advantage to a subjective judgment of greater maturity than warranted by chronological age. The purpose of the phenotype is to explore the environment and garner epigenetic experiences as well as successfully establish themselves within the adult hierarchy. How may such experiences be built? It defaults that there is an overall survival advantage for any species if the young and fit feel emboldened to participate in hunting, gathering, and protecting the family or exhibit a willingness to accept caregiver status under stress. For postadolescents to willingly accept those roles is partially dependent on endocrine status. When priming for reproduction, and as sex hormones surge, it can be hypothesized that the cellular self-assessment of maturation is enhanced based on this endocrine flow as it rises toward mature levels. In consequence, among late adolescents and teens, there is a sense of accelerated aging and the willingness to accept responsibilities that are typically deemed adult roles. It is well-established that circadian rhythms and cellular/organismal life cycles are enmeshed with the endocrine system. Therefore, if viewed within the proper cellular frame, a gap between the subjective assessment of maturity and chronological age during this developmental stage of the gradual accretion of the levels of sex hormone levels toward adult levels would be predicted. As the purpose of the phenotype is to gather epigenetic marks as environmental experience to be returned to the unicellular zygotic phase for adjudication according to first principles, the timing of the endocrine surge coinciding with the maturing organism leads to its subjective self-assessment of a level of maturity that is adequate to permit that higher level of risk away from the protection of parental oversight. Simply put, the endocrine surge enhances the cell's self-referential sense of homeostatic equipoise as it rises toward early adult levels as the "prime of life."

Adulthood is the stage of reproduction and serial accumulation of epigenetic marks. It is during this period that the major cascade of environmental stresses are experienced. Therefore, within this period, the issue of subjective age is clarified as a function of a total aggregate cellular self-referential assessment of its present equipoise based within the context of its totality of accumulated epigenetic marks versus its measuring standard as its distance from optimized conformity with cellular first principles. In effect, it is the character of the epigenetic experiences as they impact the cell, as it measures itself against its intrinsic standards that count most. It is asserted that the totality of those epigenetic impacts is measured by self-referential cells versus their own "sense" of cellular equipoise as assessed vis-a-vis first principles, which relate to perpetual unicellular roots. When cellular reserves are measured beyond the standard, homeostatic equipoise is judged in a "positive" cellular frame, which then aggregates across the multicellular organism to strike out organismal subjective senses, and is viewed as "younger" than chronological age. If adult life has been harsh, with periods of starvation, deprivation, insecurities of many types, loss of loved ones, repeated trauma, or life-threatening infectious events, then the cell's sense of equipoise degrades and is felt to be below the reference standard and is subjectively sensed by the organism as a whole as being older than chronological age.

The selection advantage of subjective age in "early life" is clear, androgens promoting the risk-taking that characterizes phenotypic agency for collecting epigenetic marks and increases the likelihood of acceptance within adult society. However, in later life the selective advantage of subjective age is harder to discern given that it occurs beyond the reproductive stage of life. Humans are outliers when it comes to longevity and the integration of older adults within the group's social structure. However, a growing body of evidence from hunter-gatherer populations suggests that humans are unlike any other species, including other closely related primates. Older males continue to hunt as well as attract and inseminate females, and older women, including those past menopause, continue to participate in the day-to-day activities of the group, including gathering and child-rearing – often referred to as the "grandmother hypothesis." Collectively, the participation of older adults creates significant survival advantages to both the individual and the group (Hawkes 2004). Thus, the epigenetic advantage for aging adults to self-perceive and act as if they were younger than their chronological age would have been selected for.

Thus, cellular self-referential appraisal is compared versus an internal reference point, which is a combination of its genomic endowment through time-clock genes, and its basal attachment to first principles. A linkage between aging and the accumulation of epigenetic experiences has been previously established. Further, the disruption of cellular time-clocks by the epigenetic modification of genes by environmental experiences that impact mTOR complexes and nutrient sensors has also been linked to the aging process (Johnson 2018). Such genomic time-clocks have been documented in many tissue types that regulate cell cycles and growth, for example, the circadian clock genes that participate in non-circadian cyclical hair growth (Geyfman and Andersen 2010). Further yet, recent research has uncovered biomarkers of aging based on DNA methylation data, which permits accurate aging for any tissues of the body across the entire life span (Horvath and Raj 2018).

In this schema, an individual's sense of feeling younger than their chronological age becomes a function of aggregate cellular background stress, which translates through cell-cell communication to become our "subjective" sensibility at the level of our minds. The cellular senome, as the cellular apparatus that permits a cell to respond to the conflicting and ambiguous environmental cues that it receives, measures a differential between an actual real-time cellular assessment of its living experience, measuring it against its own intrinsic reference standard. In general terms, if life has treated you roughly, you feel "old beyond your years" as a result of accumulated epigenetic stress and the concomitant degradation of essential cellular homeostatic equipoise, cell-cell communications, and immune status. Importantly though, any such self-referential assessment implies a measurement, and any such measurement necessitates an internal reference system. Therefore, emphasis is placed on the primacy of a cellular set of first principles of physiology embedded in cellular memory as the means by which self-referential cells can judge their current status versus a perpetual normative standard.

The advantage of this framework is that it rationalizes a number of well-documented research findings. A DNA methylation-based "epigenetic clock" (Field et al. 2018) has been identified that has strong correlates with chronological age and biomarkers of physical and mental fitness. Discrepancies between subjective age and chronological age have been attributed to "DNA methylation acceleration" which has been applied to a number of clinical conditions and has been imputed as an independent heritable trait that might be an independent predictor of mortality (Svane et al. 2018).

Although both telomere length (Bergsma and Rogaeva 2020) and DNA methylation (Feng and Lazar 2012) have been proposed as means of measuring biologic clocks, studies have confirmed that they are independent of one another. However, other studies confirm that epigenetic age acceleration is associated with clinically apparent, age-related phenotypic changes. All of these disparities rationalize within a cellular-molecular framework in which epigenetic changes are an ongoing, real-time reaction to environmental stresses that must be assessed compared to an inherent cellular standard.

Telomere length changes have been considered a possible biomarker for aging and life span. As this is considered a largely genetic endowment, it would be expected to be less mutable, and indeed, that relationship to aging is still considered equivocal. Furthermore, the sensitivity of the epigenetic clock to present moment environmental stresses is well-known. For example, HIV infection is known to accelerate age according to assessment by DNA methylation levels (Moron-Lopez et al. 2021).

There is a further advantage of placing the issue of subjective age within a cellular perspective. Several clinical patterns and common observations can be reconciled within this integrated frame. For example, common expressions based on observation are explained: "they carried the weight of their years" or "they aged overnight." Most have known an individual that experiences a disruptive life crisis and ages rapidly compared to our normative expectations. This reflects an aggregate of cellular stresses, necessarily experienced by each individual cell through its senome, as assessed through self-referential measurement, and then further expressing as a whole body phenomenon. In this manner, the seemingly schismatic gulf between self-assessed physiologic aging and actual chronological age becomes an axiomatic and predictive cell-centered phenomenon representing the gap between an actual living experience and the perpetual and essential principles of cellular life. That interstice is "subjective age."

In closing, it should be noted that the relationship between adrenarche and the onset of sexual maturation is "plastic." For example, infants born small for gestational age enter adrenarche precociously, advancing their entry into puberty, sexual maturation, and senescence. One interpretation of this phenomenon is that food deprivation during development causes intrauterine growth retardation, leading to being born small for gestational age (Grev et al. 2018). The subsequent early entry into adrenarche and sexual maturation hastens the life cycle in expectation of a more food-abundant environment in the next generation. There is a precedent for this in the way that slime molds cope with food abundance, being amoeboid in plentiful conditions, whereas they revert to their sessile colonial form in low food abundance conditions (Schaap 2011). The underlying mechanism determining these two phenotypes is cyclic adenosine phosphate-mediated cell-cell signaling (O'Day et al. 2020), linking to subjective age in humans, which likewise is ultimately determined by the timing of cell-cell communications. Hence, the reproductive strategy is the proper frame for considering the phenotypic variation for subjective age.

In this context, it should be borne in mind that food deprivation during pregnancy is a popular model for metabolic syndrome – type 2 diabetes, high blood pressure, and obesity. However, when this phenomenon is understood as an evolutionary adaptation to environmental conditions, the pathophysiology becomes an epiphenomenon. But beyond that, it highlights the significance of the role of the endocrine system in determining our behaviors and how they affect epigenetic inheritance. Suffice to say that these interrelationships provide insight to the phenomenon of subjective age, acting through endocrine control of physiology to synchronize physiologic events with behaviors. That integration of organism and environment ultimately ensures fulfillment of our genetically determined life cycle. That is the focus of the chapters that follow.

Acknowledgments William B. Miller MD and John Falk PhD contributed to this Chapter.

References

F. Alonso Debreczeni, P.E. Bailey, A systematic review and meta-analysis of subjective age and the association with cognition, subjective well-being, and depression. J. Gerontol. B. Psychol. Sci. Soc. Sci. **76**, 471–482 (2021)

V. Apgar, A proposal for a new method of evaluation of the newborn infant. Curr. Res. Anesth. Analg. **32**, 260–267 (1953)

F. Baluška, W.B. Miller Jr., Senomic view of the cell: Senome versus genome. Commun. Integr. Biol. **11**, 1–9 (2018)

A. Bergland, M. Nicolaisen, K. Thorsen, Predictors of subjective age in people aged 40–79 years: A five-year follow-up study. The impact of mastery, mental and physical health. Aging Ment. Health. **18**, 653–661 (2014)

T. Bergsma, E. Rogaeva, DNA methylation clocks and their predictive capacity for aging phenotypes and healthspan. Neurosci. Insights **15**, 2633105520942221 (2020)

D. Bohm, *Wholeness and the Implicate Order* (Routledge, London, 1980)

N. Bohr, L. Rosenfeld. *Complementarity: Bedrock of the Quantal Description*. Foundations of Quantum Physics II (1933–1958). Niels Bohr Collected Works. 7. (Elsevier, 1996)

A. Collado-Rodriguez, L. MacPherson, G. Kurdziel, L.A. Rosenberg, C.W. Lejuez, The relationship between puberty and risk taking in the real world and in the laboratory. Pers. Individ. Dif. **68**, 143–148 (2014)

C. Darwin, *The Origin of Species* (John Murray, London, 1859)

F. Demayo, P. Minoo, C.G. Plopper, L. Schuger, J. Shannon, J.S. Torday, Mesenchymal-epithelial interactions in lung development and repair: Are modeling and remodeling the same process? Am. J. Physiol. Lung. Cell. Mol. Physiol. **283**, L510–L517 (2002)

D. Feng, M.A. Lazar, Clocks, metabolism, and the epigenome. Mol. Cell. **47**, 158–167 (2012)

A.E. Field, N.A. Robertson, T. Wang, A. Havas, T. Ideker, P.D. Adams, DNA methylation clocks in aging: Categories, causes, and consequences. Mol. Cell. **71**, 882–895 (2018)

M. Geyfman, B. Andersen, Clock genes, hair growth and aging. Aging (Albany NY) **2**, 122–128 (2010)

S.J. Gould, *The structure of evolutionary theory*. (Belknap Press, Cambridge 2002)

S.J. Gould, E.S. Vrba, Exaptation—A missing term in the science of form. Paleobiology **8**, 4–15 (1982)

J. Grev, M. Berg, R. Soll, Maternal probiotic supplementation for prevention of morbidity and mortality in preterm infants. Cochrane Database Syst. Rev. **12**, CD012519 (2018)

J. Groen, D.W. Deamer, A. Kros, P. Ehrenfreund, Polycyclic aromatic hydrocarbons as plausible prebiotic membrane components. Orig. Life Evol. Biosph. **42**, 295–306 (2012)

K. Hawkes, Human longevity: The grandmother effect. Nature **428**, 128–129 (2004)

S. Hawking, *A Brief History of Time* (Bantam, New York, 2011)

O. Hiort, Androgens and puberty. Best Pract. Res. Clin. Endocrinol. Metab. **16**, 31–41 (2002)

S. Horvath, K. Raj, DNA methylation-based biomarkers and the epigenetic clock theory of ageing. Nat. Rev. Genet. **19**, 371–384 (2018)

J. Jancsary, The future as an undefined and open time: A Bergsonian approach. Axiomathes **29**, 61–80 (2019)

S.C. Johnson, Nutrient sensing, signaling and ageing: The role of IGF-1 and mTOR in ageing and age-related disease. Subcell. Biochem. **90**, 49–97 (2018)

S. Kwak, H. Kim, J. Chey, Y. Youm, Feeling how old i am: Subjective age is associated with estimated brain age. Front. Aging Neurosci. **10**, 168 (2018)

M.B. Maamar, E.E. Nilsson, M.K. Skinner, Epigenetic transgenerational inheritance, gametogenesis and germline development. Biol. Reprod. **30**, ioab085 (2021)

B.S. McEwen, J.C. Wingfield, The concept of allostasis in biology and biomedicine. Horm. Behav. **43**, 2–15 (2003)

W.B. Miller Jr., J.S. Torday, Four domains: The fundamental unicell and post-Darwinian cognition-based evolution. Prog. Biophys. Mol. Biol. **140**, 49–73 (2018)

W.B. Miller Jr., J.S. Torday, F. Baluška, Biological evolution as defense of 'self'. Prog. Biophys. Mol. Biol. **142**, 54–74 (2019)

W.B. Miller Jr., F. Baluška, J.S. Torday, Cellular senomic measurements in cognition-based evolution. Prog. Biophys. Mol. Biol. **156**, 20–33 (2020)

J.A. Mohawk, C.B. Green, J.S. Takahashi, Central and peripheral circadian clocks in mammals. Annu. Rev. Neurosci. **35**, 445–462 (2012)

J.M. Montepare, M.E. Lachman, "You're only as old as you feel": Self-perceptions of age, fears of aging, and life satisfaction from adolescence to old age. Psychol. Aging **4**, 73–78 (1989)
J.E. Morley, Androgens and aging. Maturitas **38**, 61–71 (2001)
S. Moron-Lopez, V. Urrea, J. Dalmau, M. Lopez, M.C. Puertas, D. Ouchi, A. Gómez, C. Passaes, B. Mothe, C. Brander, A. Saez-Cirion, B. Clotet, M. Esteller, M. Berdasco, J. Martinez-Picado, The genome-wide methylation profile of CD4+ T cells from individuals with human immunodeficiency virus (HIV) identifies distinct patterns associated with disease progression. Clin. Infect. Dis. **72**, e256–e264 (2021)
L. Novello, P.W. Speiser, Premature adrenarche. Pediatr. Ann. **47**, e7–e11 (2018)
D.H. O'Day, S. Mathavarajah, M.A. Myre, R.J. Huber, Calmodulin-mediated events during the life cycle of the amoebozoan Dictyostelium discoideum. Biol. Rev. Camb. Philos. Soc. **95**, 472–490 (2020)
J. Piaget, *The Essential Piaget* (Basic Books, New York, 1977)
S.W. Porges, Orienting in a defensive world: Mammalian modifications of our evolutionary heritage. A polyvagal theory. Psychophysiology **32**, 301–318 (1995)
D.C. Rubin, D. Berntsen, People over forty feel 20% younger than their age: Subjective age across the lifespan. Psychon. Bull. Rev. **13**, 776–780 (2006)
E. Scerri, *The Periodic Table: Its Story and Significance* (Oxford University Press, Oxford, 2019)
P. Schaap, Evolutionary crossroads in developmental biology: Dictyostelium discoideum. Development **138**, 387–396 (2011)
J.P. Schrum, T.F. Zhu, J.W. Szostak, The origins of cellular life. Cold. Spring Harb. Perspect. Biol. **2**, a002212 (2010)
S. Shinan-Altman, P. Werner, Subjective age and its correlates among middle-aged and older adults. Int. J. Aging Hum. Dev. **88**, 3–21 (2019)
A.M. Svane, M. Soerensen, J. Lund, Q. Tan, J. Jylhävä, Y. Wang, N.L. Pedersen, S. Hägg, B. Debrabant, I.J. Deary, K. Christensen, L. Christiansen, J.B. Hjelmborg, DNA methylation and all-cause mortality in middle-aged and elderly danish twins. Genes (Basel) **9**, 78 (2018)
J.S. Torday, A central theory of biology. Med. Hypotheses **85**, 49–57 (2015a)
J.S. Torday, Homeostasis as the mechanism of evolution. Biology (Basel) **4**, 573–590 (2015b)
J.S. Torday, The cell as the first niche construction. Biology (Basel) **5**, 19 (2016)
J.S. Torday, Pleiotropy, the physiologic basis for biologic fields. Prog. Biophys. Mol. Biol. **136**, 37–39 (2018)
J.S. Torday, The singularity of nature. Prog. Biophys. Mol. Biol. **142**, d23–d31 (2019)
J.S. Torday, W.B. Miller, Phenotype as agent for epigenetic inheritance. Biology (Basel) **5**, 30 (2016)
J.S. Torday, W.B. Miller Jr., Biologic relativity: Who is the observer and what is observed? Prog. Biophys. Mol. Biol. **121**, 29–34 (2016)
J.S. Torday, W.B. Miller Jr., The resolution of ambiguity as the basis for life: A cellular bridge between Western reductionism and Eastern holism. Prog. Biophys. Mol. Biol. **131**, 288–297 (2017)
J.S. Torday, W.B. Miller Jr., Unitary physiology. Compr. Physiol. **8**, 761–771 (2018)
J.S. Torday, V.K. Rehan, The evolutionary continuum from lung development to homeostasis and repair. Am. J. Physiol. Lung Cell. Mol. Physiol. **292**, L608–L611 (2007)
J.S. Torday, V.K. Rehan, Lung evolution as a cipher for physiology. Physiol. Genomics **38**, 1–6 (2009)
J.S. Torday, V.K. Rehan, *Evolutionary Biology, Cell-Cell Communication and Complex Disease* (Wiley, Hoboken, 2012)
M.J. West-Eberhard, *Developmental Plasticity* (Oxford University Press, Oxford, 2003)
M. Wettstein, S.M. Spuling, A. Cengia, S. Nowossadeck, Feeling younger as a stress buffer: Subjective age moderates the effect of perceived stress on change in functional health. Psychol. Aging **36**, 322–337 (2021)
L. Wolpert, Gastrulation and the evolution of development. Dev. Suppl. **116**, 7–13 (1992)
L. Wolpert, C. Vicente, An interview with Lewis Wolpert. Development **142**, 2547–2548 (2015)

Chapter 2
Superposition of Phylogeny and Ontogeny as a Quantum Mechanical Coherent Wave Collapse

Check for updates

Introduction

Confusion about our place in the cosmos is exemplified by the anthropic principle, which we are in it, not of it, first voiced by physicist Brandon Carter in 1974. The following is a cohesive way of understanding that our being is integral with the cosmos.

In Lee Smolin's Einstein's Unfinished Revolution (2019), he refers to the superposition of cohering waves, resulting in wave collapse. By homology, ontogeny and phylogeny are waves generated by cell-cell communication mediated by soluble growth factors and their cognate receptors. Consequently, the superposition of phylogeny on ontogeny results in their collapse, which can be seen as evolution (Torday and Rehan 2007). This concept represents the merging of the principles of quantum mechanics (QM) with evolutionary biology or all of biology as one continuous process. So there is no longer mere speculation about the interrelationship of QM and evolution; there is a specific set of biologic properties that are equivalent to QM.

For example, the Pauli exclusion principle stipulates that no two electrons in an atom can have the same values of the four quantum numbers: n, the principal quantum number; l, the azimuthal quantum number; $m_{\ell,}$ the magnetic quantum number; and $m_{s,}$ the spin quantum number. The first three quantum numbers are determined, whereas the fourth is probabilistic, based on time. Similarly, the cell is founded on the first principles of physiology, negative entropy, chemiosmosis, and homeostasis. Here again, negentropy and chemiosmosis are determined, whereas homeostasis is probabilistic. As such, the atom is a fractal of everything in the cosmos, and the cell is a fractal of biology. The cell is derived from the atom through the process of endosymbiosis, the endogenization and compartmentation of factors in the environment presenting as existential threats, culminating in physiology (Torday and Rehan, 2004).

J. S. Torday, *Hormones and Reality*, https://doi.org/10.1007/978-3-030-93691-4_2

Niche Construction as the Basis for the Relationship of the Cell with its Ecosystem

The relationship between biology and its environment emanates from the cell as the first niche construction (Torday 2016), the unicell endogenizing factors in its environment that posed an existential threat, forming the continuum from the outside of the cell to its interior, or Bernard's *milieu interieur*. The iterative unfolding and enfolding of quantum mechanics is infinite space and energy out of which matter can unfold, as the explicate, and enfold as the implicate, together acting as an undivided whole (Bohm 1980) from the unicell to Gaia as one continuous network.

Yet another QM feature of life is non-locality. There are genes which exhibit pleiotropy or the expression of the same gene in different tissues and organs. Such pleiotropic genes act in harmony with one another allostatically, particularly under stressful conditions. When this synchronization of genes is productive, it forms what Abraham Maslow termed "peak experiences" (Maslow 1998). However, when the individual is overwhelmed by such stress, he/she reverts to the "fight-or-flight" mode.

Quantum Mechanics Aligns with Physiology

But why should there be such alignment of quantum mechanics and physiology? Elsewhere, it has been proposed that the origin of life was due to lipids emanating from the same frozen snowball-like asteroids that delivered water to earth once it cooled down about 100 million years after it formed. When lipids are immersed in water, they float at the surface and align perpendicularly to it, with their negative hydrophilic ends facing downward into the water and their positive hydrophobic ends facing skyward because lipids are amphiphiles. Electromagnetic waves originating from pulsars or photons from the sun would have impacted on these lipid molecules, causing them to pulsate. That would have caused the formation of micelles, semipermeable protocells. The pulsing of the micelles would have caused the quantum uptake of calcium ions from the surrounding water, followed by the osmotic uptake of water molecules; intracellular calcium would have been expelled, followed by water molecules to maintain the homeostatic balance of the cell. This would have formed the basis for the quantum pulsatility of the heart, intestines, and hormone secretion, for example. Perhaps more importantly, calcium determines the "state" of the cell as homeostatic, meiotic, or mitotic.

Suffice it to say that prior to these events there was only the implicate order; it was the formation of life that gave rise to the explicate order.

Experimental Evidence for the Integration of Quantum Mechanics and Biology

Parathyroid hormone-related protein (PTHrP) is known to be a mechanotransducer, mediating the effects of gravity on the lung, kidney, and bone. More importantly, when exposed to microgravity, the intact organism and the cells that compose it are affected by the absence of gravity. In the case of lung and bone cells, they lose their evolved physiologic properties, reverting to an earlier stage in their evolutionary history (Torday 2003). In the case of yeast, they lose both their ability to orient to their environment and their capacity to reproduce (Purevdorj-Gage et al. 2006). These effects of gravity reflect the very earliest orientation of life to earth's gravity and to the electromagnetic waves alluded to above.

The wider ramifications of this relationship between gravity and the physiologic state of the cell are reflected by the role of the target of rapamycin (TOR) in cellular physiology. The TOR gene is directly "servo-ed" to the cytoskeleton of the cell (Sarbassov et al. 2004), which determines whether the cell is homeostatic, mitotic, or meiotic. The TOR gene regulates all of these facets of cellular structure and function, providing a genetic mechanism for its myriad roles in cellular life. The same holds true for plants, which unlike animals orient themselves downward into the ground as a gravitropism. According to Frantisek Balushka et al., the consciousness of plants dwells in their roots (2009), in contrast to animals, in which consciousness is localized to their skin and central nervous system (Holland 2003).

Quantum Mechanics as Physiology

The major principles of quantum mechanics have been extrapolated to cell biology – Pauli exclusion principle, non-locality, coherence, and wave-collapse – based on the common origins of both. This has been achieved using a diachronic approach, cutting across space-time, factoring out the material aspects of biology, leaving the energy flow between cells for structure, function, and homeostasis as the only remaining property of ontogeny and phylogeny. It is only in the diachronic approach to the history of the organism that its true underlying nature is revealed. This results from the fundamental way in which life copes with the ever-changing environment by eliciting genetic traits used in the past for other existential threats, referred to by Gould and Vrba as exaptations (1982). It is this self-referential, self-organizing, self-authorship that distinguishes life from nonlife. The connections are made not by random mutations as Darwin would have us think, but through interactions between the organism and its environment, causing stress that disrupts the homeostatic nilpotency of the cells involved, functionally dissociating them. The response of the cells is to produce reactive oxygen species, known to cause gene mutations and duplications, not Darwinian willy-nilly randomness, but within the context of the structures and functions involved, constrained by homeostasis.

The cells reestablish homeostatic balance over time by remodeling themselves to accommodate the prevailing, ever-changing environmental conditions. This is the cellular perspective on what evolution constitutes (Torday and Rehan 2012). The stepwise cell-cell give-and-take between the cells involved as they reconfigure themselves to fit with the conditions is quantal, not graded as Darwin thought or "punctuated" as Eldredge and Gould would have us think. The two latter approaches are descriptions, whereas the cell-cell communication process is empiric in nature. That level of resolution only became available once cells could be isolated in cell culture and studied isolated from one another and then experimentally put back together, leading to the discovery of soluble growth factors and their cognate receptors.

The Cell Is the Measure

The ancient Greeks believed that "man is the measure of all things." But now, with the recognition that there is a continuum between quantum mechanics and the cell, perhaps the dictum should be "the cell is the measure of all things." It has been proposed that the origin of life on earth began with the spontaneous formation of micelles from the lipids that accompanied those frozen snowball-like asteroids that formed the ocean that initially covered the earth. It was that instantiation that delineated David Bohm's explicate order from the preexisting implicate order. If this is correct, then in our deliberations about the role of physics in cosmology and how it impacts life should discriminate hierarchical relationships.

Nowhere is this perspective more relevant than in our understanding of what consciousness constitutes. We have been deliberating such questions formally since the time of the ancient Greeks, but they have come to a head with the psychologist David Chalmers asking the "hard question" – why we see red when we whack our thumb with a hammer? (1995) – and Clark and Chalmers formulating the "extended mind" (1998), transcending the corporeal, entering the environment. It is these ways of thinking that challenge our fundamental understanding of our orientation toward our environment that is being addressed herein. Think of it like the difference between Thomas Nagle's classic "What it is like to be a bat" versus the Vulcan Mr. Spock in the television series "Star Trek." In Nagle's assessment of consciousness, it is a function of the subjectively perceived reality of any given organism. In contrast to that, the fictional Vulcan is objectively conscious of and in sync with the cosmos itself; Captain Kirk's consciousness is primarily a function of his subjectively evolved psyche. Consequently, Kirk makes decisions influenced by irrational emotions, whereas Spock makes rational decisions based on the principles of the cosmos.

Quantum Mechanics, Evolution, and Consciousness

It has previously been claimed that there is a continuum from physics to consciousness via cell-cell signaling as physiology (Torday 2020). Moreover, there are homologies between quantum mechanical features such as Pauli exclusion principle, Heisenberg uncertainty principle, non-localization, coherence, and wave collapse and the first principles of physiology, further extending the causal relationships between quantum physics and evolutionary biology as consciousness. The present invocation of ontogeny and phylogeny as superposition for wave collapse is the clearest exposition of quantum physics as the basis for the totality of the cosmos as a singularity, allowing for the merging of our individual consciousness and the consciousness of the cosmos.

Quantum Decoherence

The usual reason for dismissing a role for quantum mechanics in cell physiology is that classical physics would cause the former to decohere or dissociate. However, we commit a systematic error by thinking about evolution from its ends instead of its means. Such after the fact reasoning is illogical, yet we continue to do so. Nowhere is that more evident than in the case of quantum versus conventional Newtonian mechanics. It has been stipulated elsewhere that the atom and the cell are homologues, i.e., of the same origin, and that in both cases they are deterministic and probabilistic. What is lacking is the realization that the cells signal to one another, acting to coordinate the quantum characteristics such that they appear consistent with classical mechanics, just as atoms do. Therefore, the quantum aspect of cell physiology does not decohere unless cellular homeostasis is disrupted, causing dissociation of the cells from one another. It is the very nature of the cell referencing the singularity at the quantum level that allows for reproduction, development, physiology, injury-repair, and evolution alike.

Discussion

In the same sense that all cells comply with the first principles of physiology, so too do they comply with QM. But these are not "one-to-one" synchronic relationships because the consequences of these fundamental principles are derived from the diachronic endogenization of factors in the environment over the course of the history of the organism. The contemporary effects of the environment are on the end products of a chain of events. For example, the effects of air pollutants on the alveoli of the lung interfere with the cell-cell communications that homeostatically control ventilation-perfusion matching. Consequently, the alveoli "simplify," reverting back to earlier stages of their evolution. Other tissues and organs exhibit the same phenomenon, which is conventionally seen in the synchronic context of disease. However, in the guise of evolution, it is realized that what is occurring is that the organism has found a way of remaining functional until it can transfer its genetic heritage to the next generation. Using this strategy, the sequential acquisition of novel, evolutionary genetic traits remains viable.

The application of the principle of terminal addition to phantom limb syndrome is representative of the diachronic approach (Torday and Miller Jr 2018). The sensing of a limb that is no longer attached to the individual in question runs counter to the efficiency of nature; what selection advantage would there be to feeling a missing limb? But when one considers this phenomenon in the context of terminal addition – the literal sequential adding on of features to an evolved trait – mediated by cell-cell communication, it would be inefficient to add new traits somewhere in the middle, or at the beginning, given that the cell-cell communications represent the history of the organism, and how such sequences interdigitate with other such cell-cell communications. If the individual in question did not have such "tinglings," everything upstream of the missing limb would go fallow, dissociating the organism from its environmental cues.

Such considerations are particularly important when considering the relationship between our individual consciousnesses and the consciousness of the cosmos. Given that the cell functions based on QM principles, as does the cosmos, and everything in between, there is a common denominator for the totality. This realization offers the opportunity for the fulfillment of Sir Thomas More's Utopia, L.L. Whyte's Unitary Principle in Physics and Biology, and E.O. Wilson's dream of Consilience.

References

F. Baluska, S. Mancuso, D. Volkmann, P.W. Barlow, The 'root-brain' hypothesis of Charles and Francis Darwin: Revival after more than 125 years. Plant Signal. Behav. **4**, 1121–1127 (2009)

D. Bohm, *Wholeness and the Implicate Order* (Routledge, London, 1980)

D. Chalmers, Facing up to the problem of consciousness. J. Conscious. Stud. **2**, 200–219 (1995)

A. Clark, D. Chalmers, The extended mind. Analysis **58**, 7–19 (1998)

S.J. Gould, E.S. Vrba, Exaptation- a missing term in the science form. Paleobiology **8**, 4–15 (1982)

N.D. Holland, Early central nervous system evolution: An era of skin brains? Nat. Rev. Neurosci. **4**, 617–627 (2003)

A.H. Maslow, *Toward a Psychology of Being* (Wiley, Hoboken, 1998)

B. Purevdorj-Gage, K.B. Sheehan, L.E. Hyman, Effects of low-shear modeled microgravity on cell function, gene expression, and phenotype in Saccharomyces cerevisiae. Appl. Environ. Microbiol. **72**, 4569–4575 (2006)

D.D. Sarbassov, S.M. Ali, D.H. Kim, D.A. Guertin, R.R. Latek, H. Erdjument-Bromage, P. Tempst, D.M. Sabatini, Rictor, a novel binding partner of mTOR, defines a rapamycin-insensitive and raptor-independent pathway that regulates the cytoskeleton. Curr. Biol. **14**, 1296–1302 (2004)

L. Smolin, *Einstein's Unfinished Revolution* (Penguin Books, New York, 2019)

J.S. Torday, Parathyroid hormone-related protein is a gravisensor in lung and bone cell biology. Adv. Space Res. **32**, 1569–1576 (2003)

J.S. Torday, The cell as the first niche construction. Biology (Basel) **5**, 19 (2016)

J.S. Torday, Consciousness, redux. Med. Hypotheses **140**, 109674 (2020)
J.S. Torday, W.B. Miller Jr., Terminal addition in a cellular world. Prog. Biophys. Mol. Biol. **135**, 1–10 (2018)
J.S. Torday, V.K. Rehan, Deconvoluting lung evolution using functional/comparative genomics. Am. J. Respir. Cell Mol. Biol. **31**, 8–12 (2004)
J.S. Torday, V.K. Rehan, The evolutionary continuum from lung development to homeostasis and repair. Am. J. Physiol. Lung Cell. Mol. Physiol. **292**, L608–L611 (2007)
J.S. Torday, V.K. Rehan, *Evolutionary Biology, Cell-Cell Communication and Complex Disease* (Wiley, Hoboken, 2012)

Chapter 3
The Periodic Table and Evolutionary Biology Are on the Vector of the Big Bang

Introduction

There is a general "sense" that there is an ultimate truth contained within the cosmos that David Bohm called the implicate order. We tend to express this thought in many different ways in both science and the humanities, yet it remains elusive. And yet there are scientific breakthroughs based on empiricism that continue to encourage us to seek that ultimate truth, such as heliocentrism, the periodic table of elements, evolutionary biology, quantum mechanics, and the Big Bang. And if there were some common origin from which all of these properties have emerged, we should hypothetically be able to identify it by finding self-similar patterns in each. It is the identification of such patterns in the periodic table of elements (Scerri 2019) and evolutionary biology (Torday and Rehan 2017) that form the basis for this article. These observations were revealed by an empiric approach to the evolution of physiology based on principles of cell-cell communication mediated by soluble growth factors and their receptors (Torday and Rehan 2012; Torday and Rehan 2017). As such, this approach is unique among the many ways in which evolutionary biology has been addressed since it is totally based on empiric evidence and is therefore faithful to Popperian refutability.

The Periodic Table of Elements and You

Eric Scerri has written extensively about Mendeleev's periodic table of elements. In his book (2019) Scerri importantly informs us that Mendeleev did not merely arrange the elements based on their atomic number, and he also used how they reacted chemically to produce specific salts to guide his decisions in constructing his table, taking into consideration variations like isotopic forms, reinforcing the

J. S. Torday, *Hormones and Reality*, https://doi.org/10.1007/978-3-030-93691-4_3

diachronic across – space-time use of atomic number to arrange the table. Atomic mass is a function of the number of protons in the nuclei of the elements, referencing the Big Bang moving forward to generate the cosmos.

Such use of empiric knowledge is akin to the way that evolutionary biology has been reverse-engineered based on cell-cell communication mechanisms in developmental biology (Torday and Rehan 2012; Torday and Rehan 2017), the three primary germ lines – endoderm, ectoderm, and mesoderm – interacting with one another sequentially, beginning with the fertilized egg or zygote, ultimately giving rise to the offspring. The premise is that embryologic mechanisms are the only way we currently know of for generating form and function biologically. By superimposing the developmental cell-cell signaling mechanisms on phylogenetic changes in phenotype (Torday and Rehan 2012), the underlying mechanisms of change from the swim bladder to the lung or the glomus of the fish kidney to the glomerulus of land animals emerge. And like the empiric details of the periodic table mentioned above, the hormonal effects on evolution provide the dynamic aspects of phenotypic change both ontogenetically and phylogenetically.

These are conventionally thought of as time-based processes, but since the physicists now tell us that time does not actually exist, these properties of biology all exclusively reference space.

It is interesting to recall here the speculation of William Crookes in his 1886 Presidential Address to the Chemical Section of the British Association that a spiral representation of the periodic table could be explained in terms of a progressive evolutionary genesis of the elements as a result of two forces, one "operating in accordance with a continuous fall of temperature" and the other showing a sinusoidal variation (simple harmonic oscillator), connected with the electric force, together producing a (double helical) generation of elements of increasing atomic mass, but periodically similar chemical properties (Scerri 2019). This speculation, which predated the discovery of basic atomic structure, the electron, the proton, and the theory of stellar genesis, which predicted isotopes and accommodated the (undiscovered) inert gases, could be seen now as exemplifying the characteristic actions of nature when operating according to the (then unknown) universal "rewrite system" devised by Peter Rowlands in his book on The Foundations of Physical Law (2014).

Thinking about how the dual forces of decreasing entropy and the sinusoidal oscillations/electrical force yielding a double helical generation of elements would coincide with the biology, the oscillations being reminiscent of the oscillating levels of oxygen in the atmosphere over the last 500 million years, fluctuating between 15 and 35% (Berner 1999). The documented periodic increases in oxygen caused the widely recognized phenomenon of "gigantism," whereas the physiologic effects of the periodic decreases are not addressed anywhere in the scientific literature, other than what has been hypothesized regarding the evolution of endothermy/homeothermy (Torday 2015). Hypoxia is the most pernicious natural physiologic stressor known. That correlates with specific physiologic changes that occurred in the pituitary-adrenal axis during this same epoch – the appearance of the parathyroid hormone-related protein (PTHrP) gene in both the anterior pituitary and adrenal cortex. The structural-functional effect of PTHrP is seen in the capillary arcades of the adrenal medulla, which expanded sometime during this era as well. That is physiologically significant because the hormonal secretions of the adrenal cortex pass down through the vascular arcades of the adrenal medulla on their way out of the adrenal to the systemic circulation. Since PTHrP stimulates the formation of capillaries (Schlüter and Piper 1998), the PTHrP produced in the adrenal cortex would have increased the vasculature of the medulla. As a result of the above, the effect of the corticoids produced in the adrenal cortex passing through the adrenal medulla, stimulating the rate-limiting step in adrenalin production, increasing adrenalin secretion by the medulla, would have been amplified by the increased surface area of the adrenal medullary vasculature (see above). Adrenalin stimulates the production of lung surfactant by the alveoli (Lawson et al. 1978), increasing their distensibility and consequently their oxygen uptake capacity. So in the aggregate, this cascade is primarily in service to alleviating the episodic hypoxia caused by the stepwise evolution of the nascent lung in adaptation to land in the short run. Over the long run, PTHrP increases the formation of alveoli (Rubin et al. 2004), constitutively increasing oxygenation.

As for why that all may have occurred, in hindsight we evolved from small shrewlike organisms that had to be nimble and quick to survive, hence the adaptive amplification of the fight-or-flight mechanism.

The Environment Gave Rise to Endothermy

All of the above have been incorporated into a "central theory of biology" for the evolution of warm-bloodedness or endothermy/homeothermy (Torday 2015). Briefly, the adrenalin that alleviated the lung alveolar constraint on oxygenation also increased the release of fatty acids from fat stores in the body. Fatty acids are the ideal substrate for metabolic production of heat, so the increased metabolic activity would have raised body temperature, ultimately becoming genetically controlled by the thermoregulatory action of oxytocin, a neuroendocrine hormone produced by the posterior pituitary gland.

The ability to maintain body temperature independently of the environment would have fostered the transition from cold- to warm-bloodedness due to more efficient metabolism since multiple isoforms are necessary for any given metabolic step in cold-blooded organisms in order to optimize the enzymatic activity at different environmental temperatures, whereas warm-blooded organisms only require one form of any given metabolic enzyme.

The increased metabolic efficiency would have facilitated bipedalism because it takes more energy to walk on two legs than on four. That led to the freeing of the forelimbs for specialized adaptations such as toolmaking and texting. Of course, that placed additional selection pressure on the brain to integrate and control such complicated physiologic properties. So that is essentially how and why our shrewlike ancestors morphed into humans.

As for the continuous increase in entropy in the environment following the Big Bang, that placed progressively greater selection pressure on biology to amplify the first principles of physiology that initially permitted negative entropy within the cell. That, in turn, would have been dependent on the lipid facilitation of oxygenation, initiated by the insertion of cholesterol into the cell membrane (Spector and Yorek 1985), followed by the evolution of peroxisomes to protect against the rising oxygen within the cell, lipids as substrate for steroid hormones, the endocrine system, and physiologic evolution.

The recognition of lung surfactant evolution as a serial preadaptation provided deep insights to the fundamental interrelationship between lipids and oxygen uptake, from lung surfactant lipids all the way back to the unicellular state based on the biosynthesis of cholesterol, the most primitive of lung surfactants; Konrad Bloch hypothesized that cholesterol was a "molecular fossil" since it took 11 atoms of oxygen to produce one molecule of cholesterol. Therefore, there had to have been enough oxygen in the atmosphere to do so, linking the oxygen and cholesterol together mechanistically in space-time. That process, in turn, gave insight to the evolution of many other physiologic traits, particularly those facilitated by parathyroid hormone-related protein (PTHrP), including those of the lung, kidney, skin, brain, and skeleton. Those insights led to a focus on the water-land transition, during which the PTHrP receptor gene duplicated, i.e., amplified.

Tiktaalik provided scientific evidence for the fossilized remains of the transition from fish to tetrapods, but there are no fossil data for the modifications of the internal organ that occurred during that process in adaptation to land since such structures would not have been preserved. However, there are extensive data for the cellular-molecular development of the organs that were essential for land adaptation, such as the lung and kidney. When such developmental mechanisms are superimposed on the phylogenetic changes, they reflect the underlying cellular-molecular changes that occurred over the course of evolution. Such hypotheses have been corroborated by gene deletions and overexpressions consistent with such evolutionary changes.

The Periodic Table and Evolutionary Biology as Diachronic Vectors of the Big Bang

The use of PTHrP signaling diachronically across space-time is homologous (of the same origin) with Mendeleev's use of chemical reactions to construct his periodic table of elements. Perhaps more importantly, receptor signaling through such "second messengers" as cyclic adenosine monophosphate and inositol phosphates gave even deeper insights to other such cell-cell signaling mechanisms occurring in tandem in other tissues and organs. The magnitude and direction of these signaling pathways are vectors for biologic change referring all the way back to the unicellular state, which dominated life on earth for 3.5 billion years. That pattern is comparable to the way in which chemical reactions guided Mendeleev's assembly of the periodic table of elements. Importantly, both the chemical and biologic reactions are diachronic mechanisms that transcend space-time. The vectoral magnitude and direction of those chemical and biologic reactions reference the singularity, from which they emerged as echoes of the Big Bang.

For example, this is why William of Ockham declared that the simplest answer is the right answer. Those observations that are most consistent with the vectors of the Big Bang present the shortest distances between two points, namely, the singularity and chemical and biologic reactions.

Mathematical expression of such vectors would lead to a way of "calculating" the value of any given property of nature. Peter Rowlands (2014), a physicist at the University of Liverpool, is seeking just such a mathematical system, arguing that there are fundamental principles of physics which underlie all that we experience. In his reduction, all of reality can be expressed in terms of "zeros and ones." We conventionally dwell on the "ones" as material reality, but it is actually zero, which Rowlands refers to as an attractor, or organizer, which is the key to understanding the fundament of physics. That is because material things are merely by-products of the energy that actually constitutes the cosmos and zero refers to the energy state.

When thought of in these terms, focusing on the cell as the basis for biologic evolution amounts to the same thing because the negative entropy within the cell is "zero" relative to the positive entropy outside of the cell. Therefore, both the animate and inanimate can be reduced to the same set of fundamental principles. Indeed, this way of thinking about cosmology is synonymous with Alfred North Whitehead's process philosophy (1978). He thought that all was energy and that occasionally matter would appear but was transient.

Whitehead thought that our focus on the material at the expense of the nonmaterial was misguided because it did not focus on relationships. In his book "Science and the Modern World," he expressed the idea that matter is "senseless, valueless, purposeless" because it is merely a transient product of the underlying energetic forces of the cosmos, a by-product. He called this perspective "scientific materialism." Whitehead found fault with the irreducible nature of matter because it masks the importance of change that nothing ever stays the same. Whitehead placed the emphasis of reality on change and that "all things flow." This concept was first voiced by Heraclitus, who said that "No man ever steps in the same river twice, for it is not the same river and he is not the same man."

Moreover, Whitehead thought that materialism masked the importance of relationships. Viewing objects as separate and distinct from all other objects was a systematic error because each object is an inert mass that is only superficially related to other things. The idea that the material is the primary state of being leads people to conclude that objects are all separated by time and space, and are not related to anything. Conversely, Whitehead thought that relationships were the primary state of being. Whitehead describes any entity being nothing more or less than the aggregate of its relationship to other entities, as the synthesis of and reaction to the world around it. Relationships are not secondary to what a thing is; they *are* what the thing is.

Whitehead's concept of reality is consistent with Mendeleev's way of assembling the periodic table. He saw elements in the context of their reactivity with other elements rather than as inert matter, with certain superficial characteristics like those described by alchemists. In that sense, it was the alchemist who determined how to make gold out of dross, not the innate characteristics of the elements that enabled them to interrelate with one another.

Similarly, by viewing organisms through the lens of their abilities to recapitulate themselves developmentally, the processes by which they fundamentally interrelate with one another comes into view. That is particularly true when development is seen in the larger context of phylogeny, offering the retrograde retracing of the process of evolution back to the unicellular state as a series of preadaptations. Such interactions between the organism and its environment over the course of evolution subordinates the material in favor of the process.

Therefore, now it can be stated that both the periodic table of elements and evolutionary biology are expressions of the process of change rather than descriptions of the superficial appearance of material objects – elements and phenotypes. Whitehead was thus correct in his perspective but had no experimental evidence to substantiate what he was espousing. Armed with evidence for this comprehensive view of reality, it behooves us to practice this philosophy.

Information Theory Meets Informatics

Information theory is the epitome of Whitehead's focus on process since it represents the mechanism for forming, storing, and sharing information. However, because it subordinates the transmission of the information to the materialism of the information, it is misguided. The consequence of that is the discipline of informatics, which is conflated with knowledge. This is syllogistic reasoning that has undermined human thought and action. The prevailing idea of informatics is that if you have not solved the problem, you just need more data. That may be true at NASA, where informatics was developed to keep track of the parts for the assembly of the space shuttle but does not apply to biologic systems – in the case of the shuttle, the data are a closed set, whereas biology is an open set, the whole being greater than the sum of its parts.

Truth Be Told

It is because of the relationship between materialism and process that we have been able to advance as a species among species. However, as David Bohm points out, we continue to exist within the explicate order, made subjective by our evolved senses, whereas there is a true reality just out of reach, referred to as the implicate order. It is because of the pseudo-reality we have formulated that we must periodically rise and then fall, ultimately succumbing to the laws of nature. Whether it is climate change or economics, we are vulnerable to our failure to comply fully with the prevailing forces of nature. We come ever closer to them as we evolve and endogenize the environment over the course of evolution, but in the current environment of the Anthropocene our narcissistic tendencies are out-stripping our natural arc for lack of insight regarding process over materialism. Once we begin engineering our heredity using CRISPR, we will deviate from our naturally evolved path toward the singularity, evolving as "silicon-based life forms" instead.

There Is only Space, There Is No Time

If, as the physicists have concluded, the only dimension is space, then how do we explain the time-based understanding of development and phylogeny? It would have to be assumed that the latter are exclusively space-filling properties of biology. In this vein, we have learned that epigenetic inheritance affects evolution, and it has been proposed that it fosters "running in place" to maintain homeostasis in consilience with the first principles of physiology, which would subsume a spatial non-temporal way of thinking about biology. This question is reminiscent of the well-documented debate between Einstein and Bergson in 1922 (Canales 2015), Einstein insisting that time is an artifact of biology and Bergson countering that time is critical for understanding any and all of biology and psychology.

If, as has been proposed, the singularity is the prototype for biology, and evolution is the process for remaining faithful to it, striving to emulate the singularity, then time would drop out of the "equation," space remaining as a point source at its minimum, the cosmos at its maximum.

A Novel Prediction of Consciousness as the Singularity

The premise for the idea that consciousness is the expression of the singularity is that there is an intersection of the singularity with physiology, the latter as the endogenization of the cosmologic environment [see Fig. 3.1]. That is to say, when the

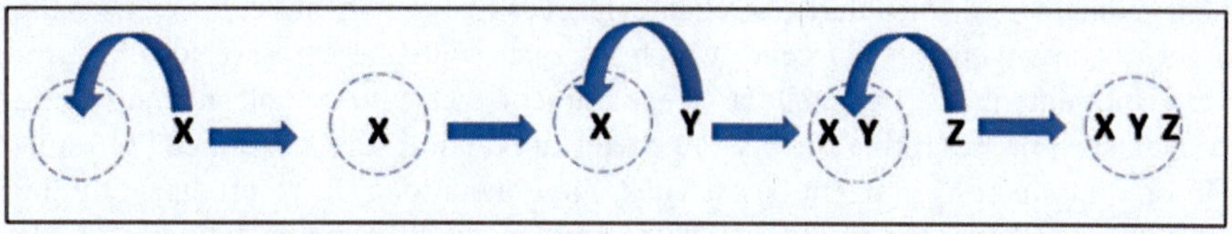

Fig. 3.1 Evolution of Consciousness. Over the course of evolution, existential threats (X,Y,Z) were assimilated, and utilized as physiologic traits. The aggregate of those traits, linked together by cell-cell communications constitutes consciousness

singularity was disrupted by the Big Bang, the information therein was fragmented but had to conform with the laws of nature. When life began on earth some 4.5 billion years ago, it too had to conform with the laws of nature. It did so by endogenizing the environment and making it useful by compartmentalizing it as physiology. In the aggregate, our physiology ascribes to the singularity as its origin, and the way in which physiology functions to maintain homeostasis is based on the same set of principles. In other words, our interoceptive sense of self is founded on the singularity as the origin of the cosmos, actualized by the physiologic principles that have evolved from the cosmos.

It is the principle of homeostasis that integrates all of these properties – when the Big Bang occurred some 13.8 billion years ago, there was an "equal and opposite reaction" based on Newton's third law of motion. That reaction is the preadaptation that we refer to as homeostasis, acting to entrain both inanimate balanced chemical reactions and life forms alike. Without homeostasis there would be no matter; there would only be energy. Alfred North Whitehead's "process philosophy" states that all is energy; matter is a transition between energy states.

The Vertical Integration of Gravity, Chemistry, and Biology as Consciousness

As mentioned above, PTHrP integrated numerous physiologic properties in vertebrates due to evolutionary selection pressure during the water-land transition. This is attributable to its mechanotransductive effects on the skeleton and lung. Such vertically integrated physiologic mechanisms offer deep insight to the interrelationships between physics, chemistry, and biology. For example, when the lung or bone cells are exposed to microgravity, PTHrP gene messenger RNA decreases rapidly, interrupting the cell-cell communication mechanisms it mediates for breathing, salt/water balance, and bone calcification, for example. In addition to being a mechanotransducer for cell-cell interactions, PTHrP is a calcium regulatory paracrine factor, acting synergistically with its effect on cell-cell communication to ensure that calcium levels in the alveolar hypophase are appropriate for optimal lung surfactant surface tension lowering function and for calcification of the skeleton in response to PTHrP signaling in bone. Even more fundamentally, when yeast are exposed to microgravity, they lose their ability to generate a calcium flux or reproduce, leaving them "comatose." Molecularly, mechanical forces like gravity affect the cell via the target of rapamycin (TOR) gene, which interacts with the cytoskeleton. In turn, matrix proteins in the extracellular space interconnect with cytoplasmic signaling mechanisms, orchestrating the myriad chemical reactions that govern cell physiology on a moment to moment basis. Thus, the gravitational force produced by the Big Bang references the singularity as the physiologic unity being emulated by the unicell, endogenized over the course of evolution, ultimately generating consciousness as the holism of being.

Biology and Chemistry as Vectoral Fractals of the Big Bang

As mentioned at the outset, the genius of Mendeleev's periodic table of elements was not in his use of atomic weight as the organizing principle but in further fine-tuning it based on the chemical reaction products that further characterized each element. This empiric approach offered a much more dynamic way of calibrating the table diachronically across space-time rather than merely arranging them based on atomic weight synchronically.

The same holds true for using the cell-cell interactions that generate form and function developmentally, applied to phylogeny to understand evolution, which is also a diachronic perspective. The value in this perspective is in being able to eliminate time from the analysis, leaving only the dimension of space to be considered, and space can be reduced to a point source without any dimensions, as pure energy, consistent with Whitehead's "process theory." In contrast to that approach, forming cladograms merely describes the progression of evolution without any underlying understanding of how or why it occurred.

Moreover, the use of such empiric data reflects the changes in the chemical and biologic reactants to their products. When shown graphically, those reactions are vectors for the magnitude and direction of change. It is hypothesized that these vectors are fractals of the Ur- vector formed by the Big Bang. And the more proximate the chemical and evolutionary biologic vectors are to the vector formed by the Big Bang, the more fundamental they are. Yet like Zeno's paradox, we can never be evolutionarily congruent with the ultimate vector of the Big Bang because we ironically must evolve in response to vectoral changes in the environment or become extinct.

This is why William of Occam's razor is indicative of the right solution, the shortest distance between two points being the simplest path.

Conclusions

In an earlier publication, it had been suggested that regressing the data for cellular-molecular lung evolution would approximate the origin of life at the intersection of the Cartesian coordinates. The present claim for that prediction is also based on cellular evolution but now as serial preadaptations, specifically identifying the singularity as the preadaptation of the unicellular state. As a "point source," the singularity would occupy no space, merely being a locale of energy essentially equaling zero space. Rowlands has similarly suggested that in math, zero is an attractor, acting as an organizing principle.

Such metaphysical ideas might help in identifying the actual nature of consciousness, which has remained an unsolved problem for thousands of years, beginning with the Ancient Greek philosophers, right up to today, when philosophers like Chalmers and Clark pose difficult questions about qualia. In the current context, it is proposed that consciousness lies at the intersection of cosmology and physiology, the product of which is what we think of as being conscious or mind.

It is offered that chemical and physical properties are vectoral fractals of the Big Bang, which could be tested by mathematically modeling key reactions for each.

References

R.A. Berner, Atmospheric oxygen over Phanerozoic time. Proc. Natl. Acad. Sci. USA **96**, 10955–109557 (1999)

J. Canales, *The physicist and the philosopher: Einstein, Bergson, and the Debate that changed our understanding of time* (Princeton University Press, Princeton, 2015)

E.E. Lawson, E.R. Brown, J.S. Torday, D.L. Madansky, H.W. Taeusch Jr., The effect of epinephrine on tracheal fluid flow and surfactant efflux in fetal sheep. Am. Rev. Respir. Dis. **118**, 1023–1026 (1978)

P. Rowlands, *The Foundations of Physical Law* (World Scientific, Singapore, 2014)

L.P. Rubin, C.S. Kovacs, M.E. De Paepe, S.W. Tsai, J.S. Torday, H.M. Kronenberg, Arrested pulmonary alveolar cytodifferentiation and defective surfactant synthesis in mice missing the gene for parathyroid hormone-related protein. Dev. Dyn. **230**, 278–289 (2004)

E. Scerri, *The Periodic Table: Its Story and Its Significance* (Oxford University Press, Oxford, 2019)

K.D. Schlüter, H.M. Piper, Cardiovascular actions of parathyroid hormone and parathyroid hormone-related peptide. Cardiovasc. Res. **37**, 34–41 (1998)

A.A. Spector, M.A. Yorek, Membrane lipid composition and cellular function. J. Lipid. Res. **26**, 1015–1035 (1985)

J.S. Torday, V.K. Rehan, *Evolutionary Biology, Cell-Cell Communication and Complex Disease* (Wiley, Hoboken, 2012)

J.S. Torday, A central theory of biology. Med. Hypotheses **85**, 49–57 (2015)

J.S. Torday, V.K. Rehan, *Evolution, the Logic of Biology* (Wiley, Hoboken, 2017)

Chapter 4
Goldilocks Effect and the Three Germ Cells or Local Paracrine Control of Homeostasis and Endocrinology

Introduction

The endocrine system is an evolved set of hormones that maintain physiologic homeostasis. As such, they monitor the moment to moment status of cells, tissues, organs, and whole organism structure and function. Like the process of cellular evolution itself, which senses when there is a loss of homeostasis and remodels itself in order to reinstate cellular-molecular "balance," the endocrine system senses such disturbances and adjusts accordingly. How the various hormonal regulatory mechanisms have evolved is not well-known, though there are a few examples that comply with developmental and phylogenetic principles.

For example, the glucocorticoid receptor evolved from the mineralocorticoid receptor as a result of the addition of three specific amino acid residues to the hormone binding site (Bridgham et al. 2006). Since this evolutionary change occurred during the water-to-land transition, it has been hypothesized that it was the result of physiologic stress relevant to the adaptive change in question. In the course of the adaptation of boney fish to land, these organisms would have experienced an increase in the force of gravity on them due to the relative difference resulting from the loss of buoyancy on land. As a consequence, there would have been a concomitant increase in blood pressure; in response, the increased shear stress on the blood vessels would have produced reactive oxygen species known to cause genetic mutations. But such mutations would have occurred in the context of cell-cell communications responsible for maintaining homeostasis. In the process of reestablishing homeostasis, the cells would remodel the structures involved and in so doing would have modified the mineralocorticoid receptors. Glucocorticoids already existed, but had no specific receptors, but because of their stimulatory effect on ß-adrenergic receptors, such modifications would have localized to those specific blood vessels experiencing shear stress. By trial and error, such vessels would have evolved into the glucocorticoid receptor through positive selection pressure for the combined

J. S. Torday, *Hormones and Reality*, https://doi.org/10.1007/978-3-030-93691-4_4

effects of the glucocorticoids offsetting both the stimulatory effect of mineralocorticoids on blood pressure and the positive effect of ß-adrenergic receptors on local regulation of blood pressure. This scenario is consistent with the duplication of ß-adrenergic receptors during the water-land transition (Aris-Brosou et al. 2009). This mechanism was particularly important in facilitating the water-land transition because it provided the basis for the regulation of the pulmonary blood pressure independent of the systemic blood pressure.

Another such example of hormonal evolution was the modification of the endostyle in elasmobranchs to form the thyroid gland in land vertebrates. The endostyle is an organ in the foregut that traps particulates during feeding. In addition, it also traps bacteria, which stimulate second messengers within the endostyle. Stimulation of cAMP in the endostyle would have caused cellular changes over the course of development and phylogeny consistent with the evolution of the thyroid gland of land vertebrates (Dremier et al. 2007).

The foregut is a plastic structure from which the thyroid, lung, and pituitary arise through Nkx2.1/TTF-1 expression. This is consistent with the concept of terminal addition, since the deuterostome gut is formed from the anus to the mouth. Furthermore, when Nkx2.1/TTF-1 is deleted in embryonic mice, the thyroid, lung, and pituitary do not form during embryogenesis, providing experimental evidence for the genetic commonality of all three organs. Their phylogenetic relationship has been traced back to amphioxus, and to cyclostomes, since the larval endostyle, the structural homolog of the thyroid gland, expresses Nkx2.1/ TTF-1.

The Phylogeny of the Thyroid

Whereas the endostyle is retained in post-metamorphic urochordates, and in adult amphioxus, the post-metamorphic lamprey has a follicular thyroid gland, which is a transformed endostyle. The existence of an endostyle in larval lampreys does not suggest direct descent of lampreys from protochordates. What it indicates is that the evolutionary history of lampreys is long and of ancient origin. Lampreys share the common feature of having filter-feeding mechanisms in their larval stage of development. Noteworthy though, the other extant agnathan, the hagfish, possesses thyroid follicles before hatching. In this vein, since hagfish evolution is considered to be conservative and hagfish history can be traced back 550 million years, this suggests that thyroid follicles could likewise be considered to have an ancient history.

Differences in the ontogeny of the thyroid gland are another example of early divergence of lampreys and hagfish during their evolutionary history. It is of interest that the method of development of thyroid follicles from a broad pharyngeal epithelium in hagfish embryos is similar to that seen during lamprey metamorphosis, when follicles arise from clumps of cells emanating from the transforming endostyle epithelium. Speculatively, maybe hagfish embryology reflects a step in the development of agnathan thyroid follicles that occurred later in lampreys, when metamorphosis appeared in their ontogeny.

The phylogenetic history of the endostyle and non-follicular thyroid tissues of vertebrates and invertebrates has previously been reviewed by Eales (Eales 1997). He focused on how exogenous compounds and peripheral mechanisms evolved to regulate thyroid status and the phylogeny of the vertebrate thyroid gland.

A key insight to thyroid-like function in a tissue is the ability to salvage iodine, and there are many non-chordate invertebrates that havė iodine-binding ability. Even though absorption of iodine compounds has not been detected in the larval lamprey intestine, yet the gut is the main site for mono-deiodination. This also seems to be the case for ascidians. Endostyles only appear in invertebrates that live in saltwater that have notochords and in freshwater larval lampreys. Though this gut adaptation may have evolved due to selection pressure favoring filter feeding, like many tissues of invertebrates, it had iodine-binding capacity, a function that developed secondarily. The iodine-binding capacity of the endostyle in ascidians and larval lampreys is well chronicled. The retention of the endostyle throughout the evolutionary history of lampreys reflects its importance in the filter-feeding apparatus and also suggests that it may have appeared when lampreys were in a pelagic marine phase.

The vertebrate thyroid gland evolved due to strong positive selection pressure for a gland that favors thyroid hormone and iodine storage during the period when ancient chordates moved from the iodine-rich marine environment to the iodine-poor freshwater habitat. Herein, the presence of an endostyle in extant larval lampreys is a reflection of their ancient marine origins, and the adult follicular thyroid gland came after the animal wanders toward freshwater. Since the adult lamprey thyroid appears post-metamorphosis of the endostyle, there was selection pressure for freshwater metamorphosis in the developing lamprey. A consequence of this metamorphosis was the transformation of the endostyle into the thyroid.

This view of lamprey thyroid ontogeny, in conjunction with the freshwater habitat of lampreys as a secondarily acquired niche, implies that metamorphosis might not have originated as a developmental strategy but occurred when lampreys moved into freshwater during their evolutionary history. Alternatively, the presence of metamorphosis, and ultimately the thyroid, in the marine environment was the primary reason why lampreys could inhabit freshwater. Endostyles in early lampreys were not required for living in saltwater since larvae with endostyles cannot tolerate even weak seawater. It is only post-metamorphosis that juveniles of some species can tolerate full-strength seawater.

The larval and reproductive intervals of the lamprey life cycle are restricted to freshwater. The capacity of juveniles of some species to osmoregulate in seawater could also be secondarily derived from the emergence of metamorphosis in the ontogeny of lampreys. Parasitic lampreys of the most ancient lineage (e.g., Petromyzontiformes) are restricted to freshwater. Why marine osmoregulation may have derived from metamorphosis may be found in modern views of the phenomenon called developmental integration.

An Evolutionary Vertical Integration of the Phylogeny and Ontogeny of the Thyroid

The increased bacterial load due to the facilitation of feeding by the endostyle may have stimulated the cyclic AMP-dependent protein kinase A, or PKA, pathway, since bacteria produce endotoxin, a potent PKA agonist (Lin et al. 2006). That cascade may have evolved into regulation of the thyroid by thyroid-stimulating hormone (TSH), since TSH affects the thyroid by way of the cAMP-dependent protein kinase A (PKA) signaling pathway. This mechanism hypothetically formed novel structures such as the thyroid, lung, and pituitary, all of which are induced by the PKA-sensitive Nkx2.1/TTF-1 pathway. Brain-lung-thyroid syndrome (Carré et al. 2009), in which infants having Nkx2.1/TTF-1 mutations develop hypotonia, hypothyroidism, and respiratory distress syndrome, is further evidence for the coevolution of the lung, thyroid, and pituitary.

The thyroid evaginates from the foregut in the mouse starting on embryonic day E8.5, 1 day before the lung and pituitary emerge, suggesting that the thyroid may have been a molecular prototype of the lung during evolution. The thyroid made iodine in the environment bioavailable by binding it to threonine to generate thyroid hormone, whereas the lung made molecular oxygen bioavailable, first by inducing fat cell-like lipofibroblasts as cytoprotectants, which then stimulated surfactant production by producing leptin, placing increased selection pressure on the blood-gas barrier by making the alveoli more compliant. In turn, that may have been further selection pressure for metabolic use of rising oxygen levels in the atmosphere and subsequent selection pressure for alveolarization, giving rise to the stretch-regulated surfactant system mediated by parathyroid hormone-related protein (PTHrP) and leptin (Torday and Rehan 2002). Next, selection pressure on the cardiopulmonary system would have facilitated liver evolution, since the progressively increasing size of the heart may have induced precocious liver development, fostering increased glucose regulation. The brain serves as a glucose "sink," and there is experimental evidence that increasing glucose during pregnancy increases the size of the developing brain. The further evolution of the brain, particularly the pituitary, would have served to further the evolution of complex physiologic systems.

The thyroid and the lung have likewise played adaptive roles during vertebrate evolution. The thyroid has mediated the use of iodine from the environment, while the lung has mediated use of the rising oxygen levels during the Phanerozoic era. These structures have accommodated the use of toxic substances for biologic purposes that have allowed vertebrates to adapt to their environment. As a result, the thyroid and the lung may have interacted cooperatively in facilitating vertebrate evolution. For example, thyroid hormone stimulates embryonic lung morphogenesis during development (Ballard 1980), while also accommodating the increased lipid metabolism needed for surfactant production by driving fatty acids into muscle to increase motility, as opposed to oxidization of circulating lipids to toxic lipoperoxides. Selection pressure for metabolism was clearly facilitated by the synergy between these foregut derivatives.

Symmorphosis as the Scientific Test of the Goldilocks Effect

In 1991 Ewald Weibel and his colleagues Taylor and Hoppeler tested the hypothesis that physiologic traits are "just right," i.e., that structure and function are economically well-matched (Weibel et al. 1991). They studied the quantitative relationships between respiration and metabolism, discovering that their hypothesis was correct, with one glaring exception – the lung itself has greater capacity than is needed physiologically. So yet again biology failed to be predictive – based on description. However, from a cellular evolutionary mechanistic standpoint, it would make sense that the lung would be "overengineered" given that it was existential for vertebrate adaptation to land and therefore there would have been excessive positive selection pressure for its evolution from the swim bladder. This perspective is reinforced by the fact that there were at least five attempts by vertebrates to adapt to land life as documented in Jennifer Clack's book Gaining Ground (2012). Such a "trial and error" effort to breech land would undoubtedly have solved the problem by overengineering itself, given that the opposite would have resulted in failure, and getting it "just right" would have required the kind of precision only achieved by engineering, not biology. Furthermore, such an evolutionary perspective on lung finds consilience with Connie Hsia's discovery that experimental lung compensatory growth is associated with increased erythropoietic receptor activity (Foster et al. 2004). At first glance, this relationship is counterintuitive, given that erythropoietin mediates the formation of red blood cells. But it is also involved in the formation of new blood vessels, so such a pleiotropic mechanism would make sense in the context of reforming alveoli after partial removal of lung tissue. The kidney is a major source of erythropoietin, referencing well-known allostatic physiologic interrelationship between the kidney and lung, best seen during fetal development when both organs contribute to the formation of amniotic fluid; failure to do so results in agenesis of both the lung and kidney in utero. In adults, the intimate relationship between the lung and kidney is seen after a myocardial infarction, secondarily causing fluid accumulation in the lung, for example.

The above interrelationships between the kidney and lung are counterintuitive at first glance. Yet the inner workings of the alveolus and glomerulus are very similar. Distension of the alveolus causes PTHrP signaling from the alveolar epithelial type 2 cell (AEC) to the lipofibroblast. The lipofibroblast then produces leptin that acts retrograde to stimulate the AEC to produce lung surfactant, preventing alveolar collapse (Torday et al. 2002). In the kidney, fluid pressure on the podocytes, the epithelial cells that line the glomeruli to produce PTHrP, binds to the mesangial fibroblasts that control fluid and electrolyte secretion into the kidney tubules (Bosch et al. 1999). Clinically, failure of one of the two structures can result in failure of the other due to disruption of such signaling mechanisms that maintain allostatic balance.

References

S. Aris-Brosou, X. Chen, S.F. Perry, T.W. Moon, Timing of the functional diversification of alpha- and beta-adrenoceptors in fish and other vertebrates. Ann. N. Y. Acad. Sci. **1163**, 343–347 (2009)

P.L. Ballard, Hormonal influences during fetal lung development. Ciba Found. Symp. **78**, 251–274 (1980)

R.J. Bosch, D. Rodríguez-Puyol, J. Bover, M. Rodríguez-Puyol, Parathyroid hormone-related protein: Roles in the glomerulus. Exp. Nephrol. **7**, 212–216 (1999)

J.T. Bridgham, S.M. Carroll, J.W. Thornton, Evolution of hormone-receptor complexity by molecular exploitation. Science **312**, 97–101 (2006)

A. Carré, G. Szinnai, M. Castanet, S. Sura-Trueba, E. Tron, I. Broutin-L'Hermite, P. Barat, C. Goizet, D. Lacombe, Five new TTF1/NKX2.1 mutations in brain-lung-thyroid syndrome: rescue y PAX8 synergism in one case. Hum. Mol. Genet. **18**, 2266–2276 (2009)

J. Clack, *Gaining Ground* (University of Indiana Press, Bloomington, 2012)

S. Dremier, M. Milenkovic, S. Blancquaert, J.E. Dumont, S.O. Døskeland, C. Maenhaut, P.P. Roger, Cyclic adenosine 3′,5′-monophosphate (cAMP)-dependent protein kinases, but not exchange proteins directly activated by cAMP (Epac), mediate thyrotropin/cAMP-dependent regulation of thyroid cells. Endocrinology **148**, 4612–4622 (2007)

J.G. Eales, Iodine metabolism and thyroid-related functions in organisms lacking thyroid follicles: Are thyroid hormones also vitamins? Proc. Soc. Exp. Biol. Med. **214**, 302–317 (1997)

D.J. Foster, O.W. Moe, C.C. Hsia, Upregulation of erythropoietin receptor during postnatal and postpneumonectomy lung growth. Am. J. Physiol. Lung Cell. Mol. Physiol. **287**, L1107–L1115 (2004)

F.Y. Lin, Y.H. Chen, J.S. Tasi, J.W. Chen, T.L. Yang, H.J. Wang, C.Y. Li, Y.L. Chen, S.J. Lin, Endotoxin induces toll-like receptor 4 expression in vascular smooth muscle cells via NADPH oxidase activation and mitogen-activated protein kinase signaling pathways. Arterioscler. Thromb. Vasc. Biol. **26**, 2630–2637 (2006)

J.S. Torday, V.K. Rehan, Stretch-stimulated surfactant synthesis is coordinated by the paracrine actions of PTHrP and leptin. Am. J. Physiol. Lung Cell. Mol. Physiol. **283**, L130–L135 (2002)

J.S. Torday, H. Sun, L. Wang, E. Torres, M.E. Sunday, L.P. Rubin, Leptin mediates the parathyroid hormone-related protein paracrine stimulation of fetal lung maturation. Am. J. Physiol. Lung Cell. Mol. Physiol. **282**, L405–L410 (2002)

E.R. Weibel, C.R. Taylor, H. Hoppeler, The concept of symmorphosis: A testable hypothesis of structure-function relationship. Proc. Natl. Acad. Sci. U. S. A. **88**, 10357–10361 (1991)

Chapter 5
Evolution of the Cell as the Flow of Energy

You could not step twice into the same rivers; for other waters are ever flowing on to you.
Heraclitus

Preface

Einstein recalled dreaming he was travelling in parallel with a beam of light at the age of 16. That vision was the basis for his "wonder year" of 1905, publishing three foundational papers on Brownian movement, photoelectric effect, and special relativity, respectively. Each of these articles was a novel perspective, acting as a "lens" through which he saw the cosmos as energy, realizing that nothing could travel faster than the speed of light.

The following is similarly a novel perspective on evolution behaving as a reduction to the flow of energy, comparable with Einstein's insights. When evolution is seen as the energetic path that takes, it becomes much simpler and therefore easier to understand.

Introduction

Darwinian evolution is predicated on the material competition between organisms, whereas cellular epigenetic evolution focuses on cell-cell communication of data from one cell to another – developmentally, phylogenetically, and as injury-repair. Such cell-cell communications are ultimately governed by the first principles of physiology, namely, negative entropy, chemiosmosis, and homeostasis (Torday and Rehan 2009). This way of framing evolution is predicated on ontogeny as the only mechanism for the formation of biologic structure and function. And ontogeny is the mechanistic basis for phylogeny as speciation. Since both ontogeny and phylogeny are based on cell-cell communications mediated by soluble growth factors and their receptors, evolution acts as one continuous, integrated, holistic process. Consequently, the material aspects of the organism can be seen as "means," not

J. S. Torday, *Hormones and Reality*, https://doi.org/10.1007/978-3-030-93691-4_5

"ends," leaving open the question as to what the "ends" of evolution are. Once the artifactual material aspects of life forms are factored out, only the ghostly flow of energy remains, within and between generations. What insights are provided by seeing evolution as the energy flows of ontology and epistemology?

The Big Bang: Vectoral Flow of All Things

The Earth is about 4 or 5 billion years old, having formed from the material aftermath of the Big Bang, which occurred about 13.8 billion years ago. The empiric evidence for that cataclysm is based on the redshift, the infrared glow detected by Penzias and Wilson in 1965. That was a paradigm shift for our understanding of the cosmos, having formed from a single source, namely, the singularity. Prior to that realization there were only anecdotal descriptions of celestial phenomena without a central theory of astronomy. Lee Smolin, an astrophysicist, has co-opted Darwinian evolution in order to explain such cosmologic phenomena as black holes and stellar evolution (1999). The existence of an "origin" for the cosmos infers that there is a vectoral "beginning," which is not unlike the origin of life as a unicell, the submersion of lipids in water spontaneously generating micelles, similarly providing a beginning for the formation of life (Fig. 5.1).

The cosmos began by disrupting the singularity, releasing energy. Homeostasis is the "equal and opposite reaction" to that event (Newton's third law of motion). Matter formed because of that homeostatic counterforce inferring that energy was the primary physical state. Based on that sequence of events, we should reconsider evolution as energy instead of matter.

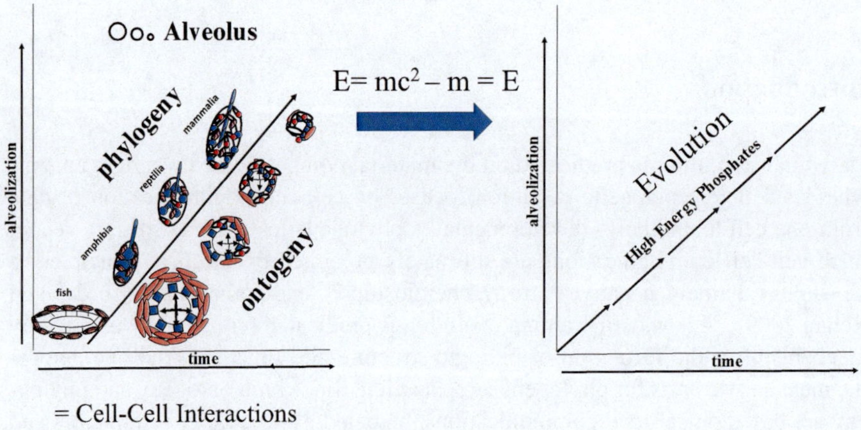

Fig. 5.1 Evolution, from matter to energy. The phenotypic evolution of lung ontogeny and phylogeny based on cell-cell communication is depicted on the left. Evolution as the flow of high energy phosphate "second messengers" is depicted on the right. Occam's razor is more consistent with the flow of energy

Complexities of Biology Are Material Artifacts of Evolution

The basis for forming complicated phenotypes are cell-cell communications mediated by soluble growth factors, binding to cognate receptors, their effects mediated by high energy phosphates as "second messengers." Such second messengers trigger the transcription of DNA within the nucleus, causing its transcription to RNA, followed by its translation to protein, or the "central dogma" of biology, the proteins stimulating the growth or differentiation of the affected cell. The target cell, in turn, makes soluble growth factors that affect neighboring cells, triggering either their growth or differentiation and so on and so forth. The tendency is to focus on the material cascade of developing forms, ignoring the unseen transfer of energy from cell to cell that causes those material changes. But by merging ontogeny with phylogeny as one unified process, eliminating the superficial spatial and temporal changes based on descriptive biology, all that remain are the energetic changes, the forms being "nodes" for the energetic "pathways." This view of development raises the question as to the origin of this energy-based pathway. Since it has been hypothesized that the unicell formed with reference to the singularity, the energy vector would appear to have been the product of the Big Bang.

Life Has Evolved Through Endosymbiosis

Unicellular life has dominated the earth for 3.5 billion years. Multicellular organisms appeared about 500 million years ago in response to bacteria effecting pseudo-multicellular forms such as biofilm and quorum sensing. But these are merely structures that form ad hoc to cope with the environment. In response to this potential threat to eukaryotic unicellular organisms, the latter evolved those cell-cell communications mediated by soluble growth factors and their cognate receptors mentioned earlier in service to multicellular organisms.

The prevailing theory of cellular evolution is symbiogenesis (Sagan 1967) – the internalization of factors in the environment that presented as existential threats. Protocells spontaneously formed from micelles with semipermeable cell membranes that mediated the acquisition of various substances. Internal membranes compartmentalized such substances and organized them biochemically as the basis for physiology.

The elements in the cosmos are ordered based on the way that stars produce the elements during the process of nucleosynthesis, beginning with the lightest element, hydrogen, progressing toward heavier and heavier elements, with more and more protons in their nuclei. Thus, this ordering of the elements forms the hierarchic basis for the laws of nature, both the cosmos and living organisms alike complying. Consequently, our consciousness is aligned with the "consciousness" of the cosmos, the cell-cell interaction mechanisms forming the infrastructure for that "logic" (Torday 2020).

"Phenotypic Agency" Addresses the Significance of Following the Energy Flows

Epigenetic inheritance is the ability of the organism to identify specific changes in the environment, assimilate them in the egg and sperm, respectively, ultimately passing such data on to the offspring. As a result, the organism gains advanced knowledge of environmental changes, enabling it to modify its phenotype accordingly. The way the offspring uses such epigenetic data is dependent on cell-cell communications mediated by high energy phosphate "second messengers," constrained by homeostasis. Therefore, the phenotype is the "agent" for the transfer of energy from one generation to the next in order to adapt to the ever-changing environment. Seen in this way, phenotypic change is a means, not an end in itself as Darwinian evolution would have us think.

Of Television Sets, Phenotypes, and Electron Flow

We have made systematic errors in our judgment based on our "explicately" evolved senses in the past, like the earth as the center of the solar system, the earth as flat, or spontaneous generation. Think about your television set, the images on the screen formed by electrons interacting with the rare earth coating on the inside of the screen, and the wires in the back of the TV that conduct the flow of electrons. It is that electron flow where the underlying principles involved emanate from. The images on the screen are like phenotypes, whereas the electron flow is the basis for the forms and sounds.

Coherence, but to What?

In his book Wholeness and the Implicate Order, David Bohm (1980) addresses how we cohere, but he does not tell us what we cohere to. Based on the precept that the cosmos is expanding, following the energetic vector formed by the Big Bang, coherence would be that vector. Deviation from the vector occurs, but that causes stress, which the organism may adjust to through auto-engineering constrained by homeostasis, but that can only occur within limits. If the stress of the deviation is too great, the organism cannot maintain its coherence, becoming extinct as a result.

Implications of Energy Flow as Evolution

William of Ockham maintained that the simplest answer is correct, yet Darwinian evolution (1859) gives rise to a complicated perspective on life – "It is interesting to contemplate a tangled bank, clothed with many plants of many kinds, with birds singing on the bushes, with various insects flitting about, and with worms crawling through the damp earth, and to reflect that these elaborately constructed forms, so different from each other, and dependent upon each other in so complex a manner, have all been produced by laws acting around us…… There is grandeur in this view of life, with its several powers, having been originally breathed by the Creator into a few forms or into one; and that, whilst this planet has gone circling on according to the fixed law of gravity, from so simple a beginning endless forms most beautiful and most wonderful have been, and are being evolved."

However, when evolution is seen as the energy paths it follows instead of being matter, as a "forest and trees" problem, it is much simpler when seen as the former than the latter (Torday and Rehan 2012, 2017). Indeed, when the principle of evolution as "serial preadaptations" or exaptations is applied, there are many places where the connections prove impossible based on material appearance, like the homology between the swim bladder of a fish and the lung of a mammal.

The analogy is between the gills of a fish and the lung of a mammal, both mediating gas exchange, as has been the case up until recently. But with the advent of functional genomics, it has become clear that the homology, i.e., being of the same origin, is that of the swim bladder as the origin of the lung. Similarly, there had been a bottleneck in tracing the evolution of the central nervous system until Nicholas Holland (2003) pointed out that the central nervous system of worms is in their skin, providing the bridge between invertebrates and vertebrates. That perspective offers deep cellular-molecular homologies between the skin and brain arising, for example, from the mechanism for the formation of the stratum corneum as a barrier against bacterial invasion from the outside and loss of fluid and electrolytes from the vasculature. The way in which lipids and host defense peptides are deposited in the skin is homologous with the mechanism for secreting lung surfactant in the alveoli and the myelination of neurons by Schwann cells. Such molecular homologies help understand why patients with neurodegenerative diseases like Tay-Sachs, Niemann-Pick, and schizophrenia have atopic dermatitis, or atopy, a skin rash, for example. And patients with asthma also develop atopy.

The genetic tie between these disparate traits is parathyroid hormone-related protein (PTHrP), which links all of these phenotypes together. The PTHrP receptor "duplicated" during the water-to-land transition (Pinheiro et al. 2012), amplifying PTHrP signaling in the lung, skin, and brain, promoting terrestrial life. Consequently, there are structural and functional cellular-molecular interrelationships between these organs.

The controversy over Darwinian gradualism versus Eldredge and Gould's "punctuated equilibrium" exemplifies this change in perspective regarding evolution. Like Bohr's explanation for light as both a particle and a wave, it is a function of how you measure it. From a molecular perspective, the same evolutionary trait occurs early on as a microscopic or submicroscopic event, but over time it becomes macroscopic, hence the differences in perspectives.

Albert Szent-Gyorgi, a founder of biochemistry, said that life is an interposition between two energy levels of an electron: the ground state and the excited state. There is experimental evidence that light can cause cell division in paramecia, for example. And the observation that a retinal cell can detect a single photon is consistent with the role of quantum mechanics in biology.

Insights from Evolution as Energy Flow

The disruption of the singularity by the Big Bang gave rise to a vectoral flow of energy that generated the cosmos, subsequently giving rise to life. That perspective naturally lends itself to the continuum from the inanimate to animate and the atom and the cell both behaving as point sources (Torday and Miller 2016a), exhibiting both deterministic and probabilistic characteristics. This reductionism lends itself to a reconsideration of evolution, particularly since epigenetic inheritance has reappeared based on the primacy of the unicellular state – epigenetic marks are biochemically incorporated into the egg and sperm (Gaydos et al. 2014) and subsequently into the offspring. As a result, we now realize that we never actually leave the unicellular state (Torday 2016), instead delegating the offspring as an "agent" for collecting epigenetic data from the environment (Torday and Miller 2016b). Seen in this light, evolutionary adaptation optimizes detection of environmental changes, emphasizing energy flow as the characteristic to be monitored.

Darwinian evolution has heavily influenced contemporary thought, from sociology to psychology, literature, and astronomy. Yet we continue to dither, seeking ways to devise a universal theory because we are the only organism that is destroying ourselves and the planet. It is hypothesized that we have focused on the material aspect of life, when in fact it is the energy side of $E=mc^2$ that holds the answer.

If, for example, literature and the arts focused on energy, perhaps they would become more consistent with the vector of the Big Bang. In that vein, instead of focusing on "supply and demand," economics should be centered on the amount of Gibbs free energy in the system, consistent with its relationship to physiology (Velev 2021). And the subject of history would be based on the free energy available to society, rather than on personalities and philosophies.

Discussion

The most compelling argument against Darwinian evolution theory is that it is illogical, founded on reasoning after the fact, based on descriptive biology. Without any alternative way of understanding evolution, Darwinism has prevailed. However, with the resurgence of epigenetic inheritance as it applies to evolution, there is an alternative way of considering evolution from its unicellular origins.

The major advances in our perception of our environment and ourselves have countered common sense perceptions thanks to empiricism – Copernican heliocentrism as the sun being the center of the solar system, the earth being round, the law of gravity. The present hypothesis that evolution is the flow of energy, not Darwin's materialistic "tangled bank," is solely founded on experimental evidence.

Our understanding of embryonic development changed dramatically with the discovery of soluble growth factor paracrine signaling for structure and function, mediated by second messengers as high energy phosphate compounds. When that mechanism was extended to phylogeny as the long-term history of speciation, it was realized that this was a solution for understanding evolution from its origins instead of its consequences. Now, seeing evolution in the forward direction, from the unicell onward, has provided insight to causal relationships starting from their physical origins in the Big Bang. Viewed in this way, as a continuous process instead of as random mutational events, has made it clear that we should have focused on the serial energy exchanges not the "endless forms most beautiful," beginning with the Big Bang.

Contemporary biology and physics are in a watershed, unable to be reduced to practice. Biomedical research is in crisis because it cannot bridge the gap between the gene and the phenotype. And quantum mechanics cannot be related to the everyday realities of Newtonian physics. It seems we have reduced these complex problems to absurdities. The underlying problem is in our after-the-fact reasoning about our beginnings and evolution. Up till now, all we have had is a compendium of organisms reduced to genes. However, genes *per se* do not form structures and functions; it is how they act within cells that brings forth those properties. This systematic error on the part of nineteenth-century biologists deciding to side with genetics in order to advance their knowledge, passing over cell biology, has mislead us to this day. And the physicists have advanced such knowledge mathematically, but now know that there is an overly reductionist dissociation of quantum mechanics and day-to-day reality.

This is a plea for continued use of the hypothesis testing scientific method in an era when scientists are defaulting to spiritualism as their only seeming alternative. For example, such initiatives as the Galileo commission and the nonduality of science, or SAND group (www.scienceandnonduality.com), opt for belief instead of rational scientific methods. Such lapses in the pursuit of scientifically based answers using hypothesis testing Popperian science is due to a lack of imagination.

References

D. Bohm, *Wholeness and the Implicate Order* (Routledge and Kegan Paul, London, 1980)

C. Darwin, *The Origin of Species* (John Murray, London, 1859)

L.J. Gaydos, W. Wang, S. Strome, H3K27me and PRC2 transmit a memory of repression across generations and during development. Science **345**, 1515 (2014)

N.D. Holland, Early central nervous system evolution: an era of skin brains? Nat. Rev. Neurosci. **4**, 617–627 (2003)

P.L. Pinheiro, J.C. Cardoso, D.M. Power, A.V. Canário, Functional characterization and evolution of PTH/PTHrP receptors: Insights from the chicken. BMC Evol. Biol. **12**, 110 (2012)

Lee SmolinSmolin, L., an astrophysicist, has co-opted Darwinian evolutionEvolution in order to explain such cosmologic phenomena as black holes and stellar evolution (1999)

L. Sagan, On the origin of mitosing cells. J. Theor. Biol. **14**, 225–274 (1967)

L. Smolin, *Life of the Cosmos* (Oxford University Press, Oxford, 1999)

J.S. Torday, The cell as the first niche construction. Biology (Basel) **5**, 19 (2016)

J.S. Torday, Consciousness redux. Med Hyp **140**, 109674 (2020)

J.S. Torday, W.B. Miller, The unicellular state as a point source in a quantum biological system. Biology (Basel) **5**, 25 (2016a)

J.S. Torday, W.B. Miller, Phenotype as agent for epigenetic inheritance. Biology (Basel) **5**, 30 (2016b)

J.S. Torday, V.K. Rehan, Lung evolution as a cipher for physiology. Physiol. Genomics **38**, 1–6 (2009)

J.S. Torday, V.K. Rehan, *Evolutionary Biology, Cell-Cell Communication and Complex Disease* (Wiley, Hoboken, 2012)

J.S. Torday, V.K. Rehan, *Evolution, the Logic of Biology* (Wiley, Hoboken, 2017)

M.V. Velev, Entropy and free-energy based interpretation of the laws of supply and demand. SN Bus Econ **1**, 1 (2021)

Chapter 6
Endothermy, Oxytocin, Vasopressin, and Civilization: A Narrative

We need a new narrative for human evolution that starts at the beginning, not at the end. The following is an effort to do so. It has previously been proposed that there is a stepwise continuum of evolved physiologic events that gave rise to human consciousness, centered on the role of bipedalism in freeing the forelimbs for toolmaking and its homologue, language, both oral and written (Torday 2021). In point of fact, it is the synergy between toolmaking and language that ultimately gave rise to Civilization.

Bipedalism was made possible by the evolution of endothermy/homeothermy, the capacity to self-regulate body heat, a physiologic trait shared only by birds and mammals, because it takes more energy to stand on two legs than it does on four. That trait emerged during the transition from water to land some 500 million years ago due to a "greenhouse" effect formed by plants generating carbon dioxide, which accumulated in the atmosphere and eventually caused a rise in the earth's ambient temperature. The rise in atmospheric temperature, in turn, partially dried up the ocean covering the surface of the earth; the rise in ocean water temperature would have partially depleted its oxygen content since the capacity of gases to dissolve in water is dependent on how cold the water is, driving boney fish, in particular, out of the water onto land.

There are two species of boney fish, physostomous and physoclistous, distinguished by the anatomy and physiology of their swim bladders, the organ that facilitates their buoyancy in water. Physoclistous fishes inflate and deflate their swim bladders by extracting gases from their circulation, whereas physostomous fishes have a pneumatic duct that connects the swim bladder to the esophagus for inflation and deflation; it is the latter type of boney fish that is our forebear, given the homology between the fish pneumatic duct and the mammalian trachea, both structurally and functionally. In fact, it has been shown that the fifty most important genes for swim bladder development in zebrafish, which are physostomous, are identical with those of the mammalian lung (Zheng et al. 2011).

Over the course of terrestrial adaptation, the evolving lung would have intermittently become inadequate to support metabolic drive, given the stepwise nature of

J. S. Torday, *Hormones and Reality*, https://doi.org/10.1007/978-3-030-93691-4_6

the cell-cell communications that determine lung evolution based on development and phylogeny. Those intermittent periods of hypoxia, which is the most potent physiologic stressor known in mammals, stimulated the hypothalamic-pituitary-adrenal axis, causing release of adrenocorticotrophic hormone from the anterior pituitary, increasing the production of cortisol by the adrenal cortex. The cortisol then passes down through the adrenal medulla, where it stimulates the rate-limiting step in adrenaline synthesis, increasing the production of adrenaline. Adrenalin then relieves the physiologic constraint on oxygenation by the alveoli, the tiny air sacs of the lung, by stimulating lung surfactant production, a soapy material that reduces the surface tension within the alveoli, allowing them to further expand, increasing their surface area for greater oxygenation. In the long run, this distention of the alveoli stimulates the production of parathyroid hormone-related protein (PTHrP) by the alveolar epithelial type 2 cells lining the air sacs, promoting the formation of more alveoli (Rubin et al. 2004).

A "side effect" of adrenaline stimulation of lung function is adrenalin dissolution of fat cells in the periphery, causing the release of free fatty acids into the circulation, where they provide the most efficient fuel for metabolism, consequently raising body temperature. The concomitant effects of increased oxygenation and metabolism are a powerful convergence of functions for the evolution of warm-bloodedness on an *ad hoc* basis. Over time, this conditional adaptation was superseded by oxytocin, a posterior pituitary neurohypophyseal hormone as the constitutively genetic regulator of body heat.

The earliest vertebrates only expressed the nonapeptide arginine vasotocin, which duplicated when jaws emerged. Such gene duplications are known to occur under physiologic stress due to the local production of reactive oxygen species by cells (Storr et al. 2013). Consequently, all jawed vertebrates now exhibit two nonapeptide forms in the brain – an oxytocin-like form and either a vasotocin or vasopressin form. The most common oxytocin-like forms are isotocin, found in bony fish, and mesotocin, found in birds, lungfish, reptiles, amphibians, and some marsupials. Cartilaginous fish have evolved at least six oxytocin-like forms, including oxytocin.

Oxytocin has a wide variety of effects on mammalian physiology, from uterine contractions in combination with milk letdown during the birthing process to various behavioral modifications that give it the nickname "the love hormone." For example, treatment of feral animals with oxytocin renders them domesticated almost instantaneously.

As hominins migrated from the Rift Valley in Africa northward some 100,000 or so years ago, they would have encountered progressively colder and colder temperatures, stimulating the production of oxytocin in order to compensate by increasing production of body heat. Perhaps it was this physiologic effect that promoted our mating with Neanderthals in the Levant, where we humans encamped for 50,000 years *en route* northward. There were also several fluctuations in the level of oxygen in the atmosphere over the course of the last 500 million years, ranging from 15 to 35%. Much has been written about the hyperoxic increases in oxygen causing gigantism, but only this author has commented on the hypothetical effect of the consequent hypoxia on the backside of the hyperoxia. Hypoxia is the most potent physiologic stressor in vertebrates. And since hypoxia is a potent stimulus for oxytocin, it may have enhanced this signaling pathway, in turn enhancing dopamine signaling in the brain, including DRD4, the "risk-taking gene," encouraging migration behaviorally, consistent with human movement northward from Africa into Europe. Polymorphisms for the dopamine 4 gene either promote or inhibit kin selection, suggesting a balance between these genes in social systems. The climate-induced increase in oxytocin would have further promoted human imagination, creativity, and ingenuity, leading to the invention of tools, including written language as a form of toolmaking.

Fast forward to the invention of the printing press in the sixteenth century, it ensured the faithful and widely disseminated publication of knowledge that fueled the generational advancement of Western civilization as we know it. And the advent of civilization acted in retrograde fashion to provide the protection and resources necessary for further imagined, creative, and ingenious activities that generated the "high" culture of northern Europe – to quote Simon Conway-Morris, "First there were bacteria, now there is New York."

The motivation for human migration to the East was somewhat different, the physiologic stress being that of having to cross wide expanses of dry, arid, dessert lands. That would have required the physiologic support of arginine vasopressin, the neurohypophyseal "partner" of oxytocin, the two hormones having evolved from a common hormonal origin, vasotocin. It is feasible that the evolution of vasopressin and oxytocin from vasotocin was due to the differential environmental stresses caused by northward versus eastward migration, particularly in light of the polymorphisms for both the oxytocin and vasotocin receptors. Whether the differential actions of these two hormones consequentially caused the differences in Western versus Eastern culture is speculative but is not beyond the realm of possibility. The fundamental difference between the two cultures is individualism in the West versus group behavior in the East, which may have been caused by these two hormones, given the differences in the behavioral characteristics caused by each.

And even the wandering aspect of human evolution may have had its origin in the stimulation of dopamine production by oxytocin within the brain. The DRD4 receptor, a dopamine D2-like G protein-coupled receptor encoded by the DRD4 gene on chromosome 11 at 11p15. 5, is known to promote wandering in humans, linking genetic control of warm-bloodedness by oxytocin with adaptation to colder climes in the northward migration and dopamine's myriad behavioral effects, both hormones playing roles in the evolution of the neocortex.

On the other hand, arginine vasopressin, also known as antidiuretic hormone, which promotes the conservation of water and electrolytes, would have facilitated the monumental migratory trek across the arid terrain from the Middle East to China, though this is pure speculation. The roles of oxytocin and vasopressin in behavior are well-recognized (Young and Flanagan-Cato, 2012), perhaps explaining differences between Eastern and Western social systems.

And then there is the "experiment of nature," the Native Americans having migrated from Asia across the land bridge to North America. Their natural behavioral tendency was to live with the land (see Braiding Sweetgrass (2020)) as the Potawatomi Nanabozhos, migrating down through the Americas from north to south.

But when Columbus discovered the North American continent, leading to the clash of European culture with Native Americans of Asian cultural origin, yet again the Europeans who migrated from Africa northward had to either destroy or dominate the Asians who had migrated eastward due to the former's penchant for individualism and concomitant ownership of land.

Perhaps significantly, Asians have more Neanderthal genes than Europeans do, suggesting that they interbred. In the book The WEIRDest People in the World (2020) the author says that kinship was the dominant force in Europe prior to the advent of the church. Perhaps that is indicative of inbreeding and the obliteration of Neanderthals on the behalf of our European forebears versus outbreeding in the case of those who migrated eastward. Was the underlying motivation for the clash differential expression of oxytocin versus vasopressin as the cause of the behaviors of each? It is well-known that Europeans tend to be more individualistic, whereas Asians are more clannish. Witness this video (https://www.youtube.com/watch?v=ZoDtoB9Abck, which documents such differences in personalities. This is consistent with the behavior of the two groups, largely due to European individualism versus Asian group behavior, in association with private ownership of property, real estate, and all.

It is of interest at this point to compare the fate of birds with that of mammals, given that both are warm-blooded and bipedal, freeing their forelimbs to fly or toolmake, respectively. As such, both avians and birds were able to wander widely across the globe, in the process collecting epigenetic "marks" that further promoted their respective evolutionary processes. And both birds and humans exhibit brain cooling during rapid eye movement sleep, raising the possibility that whatever physiologic mechanism occurs during that process is shared.

One essential property that may be fostered by all of the above is that phenotype is not engendered by superficial traits, as conventionally thought, but by its role in collecting epigenetic marks for Lamarckian inheritance, which would have been promoted by the enhanced mobility due to endothermy in combination with flight on the part of birds, and toolmaking and language, both oral and written in humans, the latter achieving equivalency with the former in the effort to collect data from the environment. The capacity for written language, particularly when converged with moveable type, led to human civilizations far beyond flocks of birds, even at the heights of migrations.

And in a global sense, it is useful at this point to recognize the exaptations of lipids in the evolution of the above behavioral characteristics (see Fig. 6.1), beginning with endothermy being initiated by the metabolism of free fatty acids as the conditional basis for endothermy, followed by bipedalism as an exaptation of lipids immersed in water, the molecules aligning themselves perpendicularly to the water's surface due to their positively and negatively charged ends, the negative ends being water soluble, leaving the positive ends pointing upward into the atmosphere. When these lipid molecules pack together, they reduce the surface tension caused by the

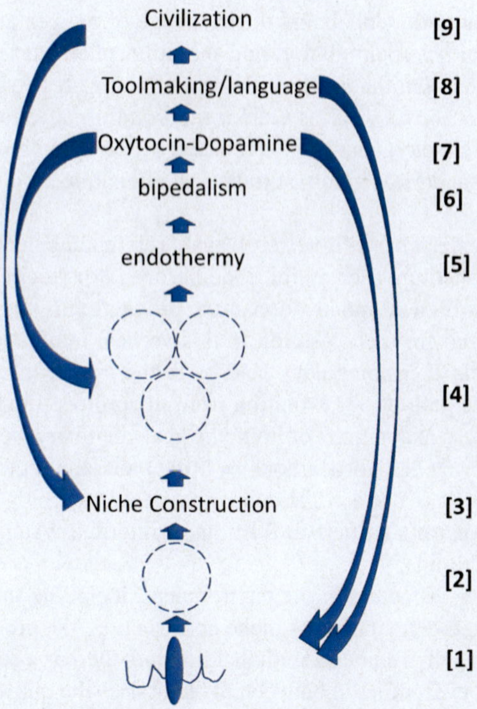

Fig. 6.1 Vertical integration of vertebrate physiology. Initially, **[1]** lipids oriented vertically to the surface; **[2]** subsequently, the lipids formed micelles; **[3]** these primitive cells endogenized factors in the environment, forming the first niche construction; **[4]** in response to prokaryotes forming biofilm and quorum sensing, eukaryotes formed multicellular organisms; **[5]** endothermy arose as a product of physiologic stress, facilitating bipedalism **[6]**; oxytocin evolved to genetically control endothermy **[7]**, stimulating dopamine; **[8]** freeing of the forelimbs due to bipedalism gave rise to toolmaking and language; synergy between toolmaking and language led to civilization **[9]**. Underlying that synergy were the various lipid-based exaptations, highlighted as blue arrows

Van der Waals forces causing the water molecules to attract one another. As a result, the lipids coalesce to form micelles, spheres with semipermeable membranes, which were the origin of cells. The subsequent existential threat posed by prokaryotes (bacteria), exhibiting pseudo-multicellular traits like biofilm and quorum sensing, led to eukaryotes (our forebears having "true" nuclei) practicing cell-cell communications that gave rise to multicellular organisms. The next milestone in vertebrate evolution was the synthesis of cholesterol and its insertion into the cell membrane, causing the cell membrane to thin, facilitating locomotion (cytoplasmic streaming) and oxygenation, which promoted metabolism. That sequence of events is what led to the evolution of the lung as the land vertebrate's means of oxygen uptake, also resulting from the facilitation of gas exchange by the production of lung surfactant, a mixture of lipids and proteins that reduces surface tension, allowing for the progressive decrease in the surface area of the alveolar air sacs, from the huge faveoli in frogs down to those of the human lung, increasing the surface area-to-blood volume ratio, enhancing the exchange of oxygen and carbon dioxide between the alveolar microcirculation and the atmosphere. It is that relationship between lipids and oxygen that became the rate-limiting step for the water-to-land transition alluded to above. And as stated, the conditional effect of physiologic stress due to inefficient oxygenation was relieved by the stimulation of cortisol production by the adrenal cortex, in turn stimulating adrenaline production by the adrenal medulla.

Cortisol is synthesized from cholesterol, again referencing the lipid basis for life. Furthermore, the constitutive basis for endothermy is oxytocin, a product of the posterior pituitary, which stimulates dopamine production by the brain, dopamine also being synthesized from cholesterol. In this vertical integration of our evolved capacity to ultimately thermoregulate, leading to the evolution of toolmaking and language, we see the path of our evolution from the protocell to higher and higher levels of physiology as exaptations of lipid physical chemistry, bearing in mind that pulsars produce polycyclic hydrocarbons (= lipids) and distribute them throughout the cosmos. Perhaps it is not a coincidence that those asteroids that delivered the frozen water to the earth's surface to form the primordial waters that covered the earth also contained lipids.

The endogenizing of factors in the environment, including that of other organisms may have been the beginning of niche construction, the process of modifying one's own immediate environment. With that in mind, there is a continuum from the unicell to Gaia, and everything in between, which raises the question as to what the origin of such a phenomenon might be? The immersion of lipids in water giving rise to life seems highly fortuitous, yet lipids have the interesting property of hysteresis or molecular "memory." Picture those lipid micelles floating at the air-water interface, being warmed by the sun by day, and cooling at night. When the micelles were warmed, they deformed and enlarged due to the partial liquification of the lipids, in the process trapping more calcium ions; when these micelles then cooled, they contained ever-increasing amounts of calcium, which is toxic to lipids, causing them to denature. Yet they survived, raising the question as to how they did so? I would speculate that there was a subset of these micelles that were able to form calcium channels that mediated the entry and exit of calcium from these protocells, producing the calcium fluxes characteristic of life. Whether it is a paramecium exposed to a drop of sugar water in its surroundings or on my tongue in an MRI, the calcium fluxes generated are indicative of consciousness.

References

J. Henrich, *The WEIRDest People in the World* (Farrar, Straus, Giroux, New York, 2020)

R.W. Kimmerer, *Braiding Sweetgrass* (Milkweed Editions, Minneapolis, 2020)

L.P. Rubin, C.S. Kovacs, M.E. De Paepe, S.W. Tsai, J.S. Torday, H.M. Kronenberg, Arrested pulmonary alveolar cytodifferentiation and defective surfactant synthesis in mice missing the gene for parathyroid hormone-related protein. Dev. Dyn. **230**, 278–289 (2004)

S.J. Storr, C.M. Woolston, Y. Zhang, S.G. Martin, Redox environment, free radical, and oxidative DNA damage. Antioxid Redox Signal. **18**, 2399–2408 (2013)

J.S. Torday, Cellular evolution of language. Prog. Biophys. Mol. Biol. **167**,140–146 (2021)

L.J. Young, L.M. Flanagan-Cato, Editorial comment: oxytocin, vasopressin and social behavior. Horm. Behav. **61**, 227–229 (2012)

W. Zheng, Z. Wang, J.E. Collins, R.M. Andrews, D. Stemple, Z. Gong, Comparative transcriptome analyses indicate molecular homology of zebrafish swimbladder and mammalian lung. PLoS One **6**, e24019 (2011)

Chapter 7
Dialectical Energism

Space and Time are the modes by which we think, not the conditions in which we live.

Einstein

Introduction

Despite the historic advances that have been made in understanding such fundamental questions as the origins of the physical realm and how and why we have evolved, there are equally perplexing paradoxes that are unexplained by our conventional way of thinking about "reality." No greater an example is there than the failure to reconcile quantum mechanics with our perception of reality. The consensus is that quantum mechanics is the ultimate way to understand physics, but it does not coincide with how we experience daily life. For example, nonlocality is the simultaneous occurrence of an event both in our immediate realm and somewhere off in the distant cosmos, which Einstein referred to as "spooky science." Or the recent discovery that Yttrium atoms can self-organize (Zhang et al. 2017) or the influence of living organisms on the function of random number generators (Gober 2020), suggesting that there is more to life than meets the eye.

To reconcile these paradoxes, we may consider the hypothesis that we misconceive of our apparent material existence as the foundational state of being, though cellular evolution is based on the flow of energy between cells, so it is the fundament of evolution as all of biology. The primacy of energy flow between cells as cell-cell communication is recognized in embryologic development, homeostasis, and injury-repair, not the material state of being that Darwinists would have us believe. The material aspect of life derives secondarily from the internalization of factors in the environment, referred by Lynn Margulis as symbiogenesis theory. Suffice to say that evolution only appears to be due to reproductive success, but that is an epiphenomenon, whereas physiologic corrections to the loss of free energy or adaptations are the actual determinants of evolution.

J. S. Torday, *Hormones and Reality*, https://doi.org/10.1007/978-3-030-93691-4_7

Rationale

Quantum mechanics is the ubiquitous mechanism underlying the cosmos, not classical physics. Given that, what biologic properties are consistent with "quantum mechanics?" Although the data are thin and scattered, fertilization of the egg by the sperm causes a calcium burst that qualifies as a quantum event (Gordon and Gordon 2016). And there are experimental data to support the direct effect of photons on cell division (Fels 2009), as well as the effect of sound on vibrotactile sensors in skin (Cobo et al. 2020). Similarly, there are data to show that the sense of smell is quantal (Brookes et al. 2012). Moreover, Radin and others have shown that the mind can affect the function of a random number generator (Mason et al. 2007). Experiments reported by Rene Peoc'h similarly demonstrate the influence of life on Turing automata (Gober 2020).

Given such evidence, the question is how and why might the effects of quantum mechanics on biology have arisen over the historic course of evolution? One possible source is the lipid micelle as the prototypical origin of the cell. It arose from the lipids generated by thermal vents in the seafloor and from pulsars in deep space. When immersed in water (Fig. 7.1), lipids will localize to the air-liquid interface, orienting themselves perpendicularly to the surface of the water due to their charged molecular "poles" as amphiphiles. In that configuration, the negatively charged ends of the lipid molecules were susceptible to electromagnetic forces and photons, rendering them like "antennae," affecting their behavior. Based on that ontology, the other such polarized orientations to the environment – the twists and turns of embryogenesis, deuterostomy, bipedalism, the vagal nerve, the bicameral cerebrum,

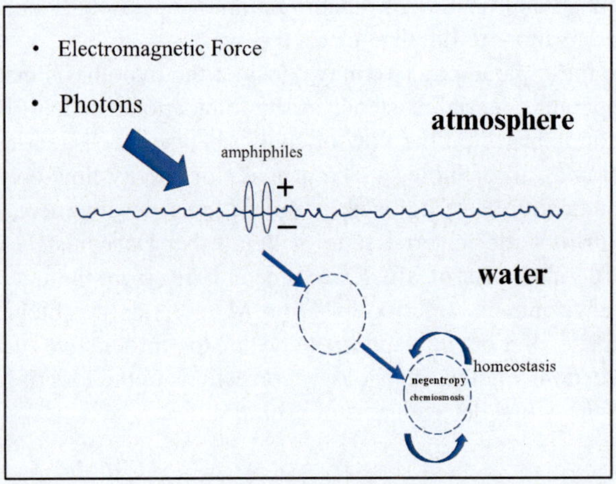

Fig. 7.1 The Origin of negentropy. Lipids are amphiphiles, with a positive and negative "pole." The lipid molecules line up perpendicular to the water surface, their negative end in the water, their positive end pointing upward into the atmosphere. Electromagnetic forces and photons impinge on the lipid molecule, affecting their behavior

and the left-right brain – have all perpetuated this antenna-like homology in attempting to mitigate against the ambiguity of life's origin (Torday and Miller 2017).

Cellular Evolution

In a series of publications, it has been proposed that by centering our focus for evolution on the unicellular state, the actual course of evolution can be understood. That is to say, for the first time we can trace the path of evolution from its origin instead of from its consequences. We know that reasoning after the fact is illogical by definition, but absent any other approach, description has been the only way we have had to pursue the most important and perplexing questions faced by humanity – how and why we exist?

The method used to deconvolute evolution, referring all the way back to the unicell, is based on what Gould and Vrba termed "exaptation" or "the process by which features acquire functions for which they were not originally adapted or selected." This principle was applied to the well-established mechanism of embryogenesis, serial cell-cell communications mediated by soluble growth factors and their receptors – the stepwise series of cellular interactions that starts with the fertilized egg and culminates with the birth of the offspring; in actuality, such cell-cell interactions do not terminate at birth. Rather, they persist until death as the homeostatic mechanisms we recognize as physiology.

Using that approach made it possible to deconvolute the evolution of the gas exchanger, beginning with the mammalian lung alveolus, working retrograde both developmentally and phylogenetically from fish to amphibians, reptiles, avians, and mammals. In that last stage of vertebrate evolution, the homology between the swim bladder and lung was elucidated (Zheng et al. 2011), revealing that the cell-cell interactions for vertebrate gas exchange actually referenced an adaptation to gravity, since the function of the swim bladder is to facilitate buoyancy using gas exchange – in other words, the mechanism of gas exchange transcends its superficial appearance, divulging the true nature of evolution in adapting to the prevailing, ever-changing environmental conditions from water to land. That realization advanced the concept of evolution from superficial adaptive structures to their underlying transcendent functions, from phenotype to process.

In that vein, it was realized that there are homologies between the atom and the cell, offering a way to understand how quantum mechanics might actually interface with biology. The denouement for that interrelationship is lateralization of the cerebral cortex, the left and right brains having opposite perceptive capacities, acting to accommodate such quantum mechanical properties as the Heisenberg uncertainty principle, coherence, entanglement, and nonlocality biologically.

The exaptive principle was further exploited by hypothesizing that the unicellular state can be reduced to the "first principles of physiology" – negative entropy, chemiosmosis, and homeostasis – which are the ultimate referents for all of life (Torday and Rehan 2009).

It should be highlighted here that life begins in an ambiguous state, perpetuated by negative entropy, fueled by chemiosmosis, and controlled by homeostasis.

Based on the exaptive approach, the first principles of physiology beg the question as to what those principles refer back to in the history of evolution. Founded on experimental evidence, it is known that the primordial ocean that covered the earth beginning about 100 million years after the formation of the planet was formed by snowball-like asteroids that freely pelted its surface due to lack of an oxidative atmosphere; those asteroids also contained lipids or polycyclic hydrocarbons produced by pulsars. When lipids are immersed in water, they spontaneously form prototypical cells or micelles. The key principle that catalyzed life was negative entropy, a concept first proposed by Erwin Schrodinger in his book *What is Life?* He thought that the unnatural energetic state of negative entropy was due to the cell utilizing energy from its environment, but that still did not explain how and why that state of being evolved. Leydesdorff et al. (Leydesdorff et al. 2018) have found that occasionally sets of mutual information can generate negative entropy. In their explanation of this phenomenon, they speculated that one of the two mutual information states entrains the entropy of the other. When lipids are immersed in water, they align perpendicularly to the surface, with their negative poles pointing down due to hydrophilia and their positive poles pointed upward into the air due to hydrophobia. When the lipid molecules do so, they pack together and reduce the surface tension of the surrounding water, effectively countering the Vander Waal's forces, producing negative entropy within the water phase.

The Cell as the Original Niche Construction: From Unicell to Gaia

Niche construction is a burgeoning concept in biology. It is based on the observation that organisms actively modify their immediate environment, optimizing their viability. This phenomenon was first documented by Darwin in *The Formation of Vegetable Mould Through the Action of Worms* (1881), in which he noted that earthworms retain their aquatic kidneys on land. Niche construction is common among organisms, as exhibited by beavers building dams or humans building homes, towns, cities, states, and nation states, culminating in Gaia or "Earth Mother," Lovelock's holistic concept of an organic earth. This idea may seem superficially descriptive at first glance, but when merged with symbiogenesis it takes on a much more fundamental meaning – this coming together of the environment and the cell forms a functionally integrated process that constitutes the global ecosystem.

The Cell as the Origin of Social Systems: Oral and Written Language, Morality, Economics, Education, and the Humanities

Now, by considering the innate properties of the cell as the basis for behaviors in the context of niche construction as ecologies, the image of society/culture/civilization takes on a holism intractable without that viewpoint being expressed above. David Bohm expressed this behavior abstractly as "enfolding and unfolding" (Fig. 7.2). Moreover, the interactions between the individual and society are synergistic, fostering ever-more intricate and dynamic social forms due to the fractal nature of the first principles of physiology.

From Dialectical Materialism to Dialectical Energism

Dialectical materialism is a philosophy of science and nature based on the work of Marx and Engels that emphasizes the importance of real-world conditions. Marx thought that material conditions were contradictory and that society could resolve them. Dialectical materialism accepts the role of evolution in providing progressive levels of existence that emerge through ontogeny. That perspective is in contrast to Hegel's dialectical that emphasizes human experience based on the mind's perception of reality.

But we systematically err in assuming that we and everything around us are materials. The default mode that everything is energy, applied to Marx's dialectical

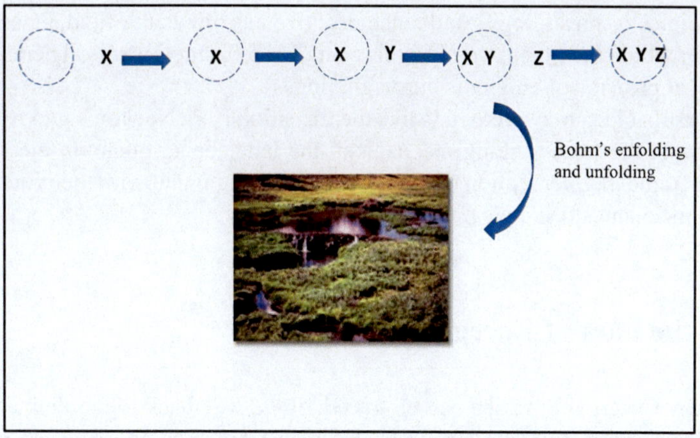

Fig. 7.2 Niche construction. The cell evolves by endogenizing environmental factors. This "enfolding" into the cell forms a continuum with the external environment by "unfolding" to generate an integrated ecologic niche

approach, would translate to dialectical energism. As such, this is Alfred North Whitehead's process philosophy.

Congruence of Quantum Mechanics and Evolution Theory

Logically, ontology and epistemology must be in sync with one another or the vectoral trajectory of the cosmos would be misaligned with biology. Endogenization as the basis for evolution is based on the homology between the internal milieu of the cell and the external environment, both operating based on the laws of nature. And the mechanism of epigenetic inheritance, the organism monitoring for change and collecting data from its surroundings accordingly, ensures harmony with the ever-changing environment.

The Necessity for a Diachronic Approach in Order to Establish the Epistemology

The question of what consciousness is reduces to ontology and epistemology. If we begin with the brain as the "seat" of consciousness, we end up with Plato's concept of "theater of the mind" or that "it is all in our heads." On the other hand, if consciousness is the "wetware" of the consciousness of the cosmos, it is the distributive property of our physiology, as determined by cell-cell communication. In order to resolve this dichotomous way of thinking about consciousness, a diachronic, across space-time approach is the solution to the problem – if you use the exaptive approach to evolution based on cell-cell communication, you end up with the unicell and its first principles of physiology. And since negative entropy is the fundamental principle upon which the other two principles act, it is a form of energy. Therefore, the hierarchical primacy of energy is our beginnings.

Alien as that idea may seem, it is like the transition from Newton's gravity theory to Einstein's, the former being mechanical, the latter being quantum mechanical. With that frame, our perception of ourselves and its alignment with the principles of quantum mechanics becomes tractable and practicable.

Life as the Flow of Energy

Negentropy fosters life, which can be traced all the way back to the delineation of the explicate order from the implicate order by reverse-engineering the cell-cell interactions common to both ontogeny and phylogeny. Prior to the appearance of the first cell, there was no explicate order. The superposition of phylogeny on ontogeny may cause a wave collapse that nullifies the material aspects of both ontogeny and phylogeny, leaving the residual high energy phosphates that mediate embryologic development, physiologic homeostasis, and phylogenetic speciation, attesting to the primary role of energy in evolution. More importantly, retracing the evolution of physiology back to its origin in the unicell allowed for the identification of the first principles of physiology, prime among them being negentropy or the energetic state of the cell. Therefore, it is the biologic premise that should define the physics and mathematics, not the other way around, as convention would dictate.

The Periodic Table as an Algorithmic Manifestation for the Role of Negentropy as an Organizing Principle

Mendeleev's version of the periodic table of elements was a breakthrough in our recognition that there are demonstrable organizing principles in the cosmos. Basically, the insight was that the number of protons in the nucleus was a common denominator with which to "normalize" the relationships between the elements (Scerri 2019). The ontology of the table derives from the stars, which produce light by converting matter into energy, in the process going through the elemental sequence from hydrogen to heavier and heavier elements.

The epistemology derives from the binary nature of the tablature, the overt synchronic appearance of the elements at one level of existence, atomic number diachronically appearing at the other level as a means of connecting the elements to the material fundament.

Using that synchronic/diachronic approach, a periodic table of biology could theoretically be constructed using the same set of principles once evolution is reduced to cell-cell communication. That is to say, the phenotype is the synchronic level of existence, whereas the developmental signaling mechanisms that are responsible for generating form and function are at the diachronic level. And since social systems are derived from biologic forms, a periodic table could also be constructed for various aspects of society – education, economics, government, morality, the humanities, aesthetics, sciences, mathematics, etc.

Left Brain-Right Brain as a Means of Perceiving Quanta

Beginning with the origin of life on earth, there has been an upward projection in vertebrate phylogeny due to the behavior of lipids in water (Fig. 7.3). As an amphiphile, the two poles of the lipid molecule are positively and negatively charged, respectively. When lipids are immersed in water, they naturally orient themselves perpendicularly to the surface of the water, their negatively charged (hydrophilic) ends pointing downward into the water and their positively charged (hydrophobic)

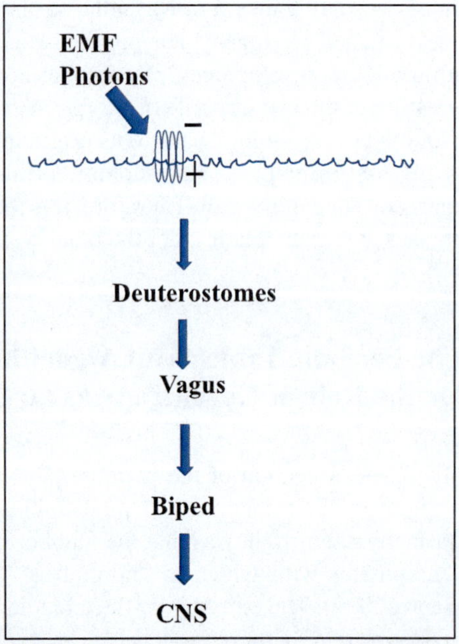

Fig. 7.3 Upward projection in vertebrate phylogeny. Starting with lipid amphiphiles aligning perpendicular to the surface of the water covering the earth to embryologic development, deuterostomy, vagal enervation, and bipedalism

ends facing upward, acting like "antennas," detecting and responding to electromagnetic forces like photons and pulsars. Subsequently, the lipids form micelles, protocells surrounded by semipermeable membranes that allow the passage of molecules in and out of themselves, due to osmotic pressure. These micelles float on the surface due to the buoyancy of lipids, where they are warmed by the sun and liquified, increasing in volume, taking up more calcium and water as a result. At night the micelles contract due to cooling, reconstituting their original shape due to hysteresis or "molecular memory."

Over time the micelles accumulate more and more calcium ions, which may become toxic, denaturing the lipids; a subpopulation of micelles evolved the ability to actively regulate the amount of calcium within themselves using ion channels. This is particularly true of neurons, which are "professional" calcium ion mediators. Fast forward to multicellular organisms, which have evolved a nervous system, foremost being the vagus, which aligns itself vertically in the body from bottom to top in accord with deuterostomy, becoming myelinated, i.e., lipid-coated, as it progresses forward developmentally and phylogenetically, interconnecting the adrenals, gut, heart, and cranial nerves.

This hypothetically was the molecular origin of bipedalism, which homologously projects upward. Parenthetically, one of the 12 cranial nerves, the hypoglossal, innervates the tongue to facilitate oral communication as a homologue for the vertical integration of cell-cell communication.

Such active orientation against the pull of gravity is attested to and underscored by the effect of microgravity on the cellular phenotype (Torday 2003). And in contrast to that, Baluska has hypothesized that plants have their "central nervous systems" in their roots, thus orienting themselves downward *toward the source of gravity*, suggesting the fundamental role of gravity in phenotypic expression.

In the aggregate, these observations regarding the effect of physical orientation on environmental awareness can be further reduced to the penultimate role of "left brain/right brain" on the quantum mechanical orientation of biology, mediated by the endocrine system.

The gyrations performed by lipid molecules from their alignment at the water-air interface to the vagus, to bipedalism, to left-right brain are reminiscent of the twists and turns of the developing embryo. In that vein, Wolpert has famously said that the most important thing you will do in your lifetime is gastrulate, referring to that critical step in embryogenesis when the mesoderm is introduced into the bilayered embryo. And consider the cerebrum itself, mostly composed of lipid-coated neurons whose tracts are indicative of their function. And to paraphrase, it is "lipids all the way down."

Experimental Evidence for the Primacy of Energy as the Level of Selection, Not Matter

From an a priori perspective, the cosmos began as a burst of energy, exploding the singularity into asymmetries resolved through balanced physical, chemical, and biologic reactions. Based on that perspective, the precedent for the reconciliation of biology mediated by evolution would also follow the sequence from energy to matter, the "equal and opposite reaction" to that explosion being homeostasis (Torday 2015).

Discussion

It is proposed that we make a systematic error in assuming that our primary state of being is material, whereas the extreme reductionist approach taken here points to energy as the actual primary state of being, matter having been derived from energy due to the "equal and opposite reaction" generated by the Big Bang. This perspective hypothetically resolves a major contemporary problem in physics – that it does not seem to correspond to our life experiences. It is hypothesized that this is due to the upside-down way in which we perceive and therefore misconstrue our being as material. On the other hand, if we saw ourselves primarily as quantum energy, quantum mechanics might be congruent with biology, physiology, medicine, sociology, psychology, economics, the Humanities, etc. As a result, these disciplines would be reconciled much the same as they were with the advent of physics and chemistry, allowing for the consilience that E.O. Wilson talked about in his book of the same title, but now based on fundamental principles rather than on associations and correlations.

The title of this article refers to Marx and Engel's *Das Kapital*, containing the economic theory of dialectical materialism. As expressed above, this theory may not be effective because it is based on the false premise that we primarily operate in a material world, when in fact we should see our relationships based primarily on energy flows.

The insight to the primacy of energy as the prevailing state of being, as expressed in this article, derives from the merging of the experimental evidence for cell-cell signaling underpinning both ontogeny and phylogeny (Torday and Rehan 2012, 2017). By tracing the cell-cell communication mechanisms from their current form backward based on Gould and Vrba's exaptations or preadaptations, evolution of the lung alveolus leads all the way back to the unicellular state and further still to the first principles of physiology, identifying negative entropy as the sine qua non of life based on Schrodinger's *What is Life?* (1944). The signaling pathways are mediated by high energy phosphate "second messengers," culminating in the entropic energy state of the cell, attesting to the primacy of energy in the evolution of physiology.

It is such empiric evidence that allows us to transcend the subjective explicate order and gain insight to the implicate order as holism. Language acts as a tool for leveraging empiric data that bridge the gap between the explicate and implicate orders, and no greater a language exists in this context than mathematics. And considering that math can express infinity as 1/0, it exists in both orders, exhibiting its power to reveal the continuum between them.

Life is only possible because of its origin in negentropy as a local energy dialectical for the backdrop of entropy in the cosmos. That dialectical dynamism must be maintained and sustained in order for the species to remain viable in the face of an ever-changing environment. This has been the case ever since the first instantiation of life as the lipids that originated in pulsars, immersed in water to spontaneously form micelles able to interact with the environment and "remember" their history. It is this latter property that is necessary for evolution to perpetuate the species, interrogating the environment for changes relevant to the experiences of the organism. Such "inhomogeneities" are detected at the quantum level, perhaps by the left-right brain, and are biochemically assimilated within the germ cells of the parent organism in anticipation of being transferred to the offspring during reproduction in order to ultimately adapt – **that is evolution, and such is life**.

References

J.C. Brookes, A.P. Horsfield, A.M. Stoneham, The swipe card model of odorant recognition. Sensors (Basel) **12**, 15709–15749 (2012)

R. Cobo, J. García-Piqueras, Y. García-Mesa, J. Feito, O. García-Suárez, J.A. Vega, Peripheral mechanobiology of touch-studies on vertebrate cutaneous sensory corpuscles. Int. J. Mol. Sci. **21**, 6221 (2020)

D. Fels, Cellular communication through light. PLoS One **4**, e5086 (2009)

M. Gober, *An End to Upside Down Thinking* (Waterside Productions, Cardiff, 2020)

N.K. Gordon, R. Gordon, *Embryogenesis Explained* (World Scientific, Singapore, 2016)

L. Leydesdorff, M.W. Johnson, I. Ivanova, Toward a calculus of redundancy: Signification, codification, and anticipation in cultural evolution. J. Assoc. Inf. Sci. Technol. **69**, 1181–1192 (2018)

L.I. Mason, R.P. Patterson, D. Radin, Exploratory study: The random number generator and group meditation. J. Sci. Explor. **21**, 295–317 (2007)

E. Scerri, *The Periodic Table: Its Story and Significance* (Oxford University Press, Oxford, 2019)

J.S. Torday, Parathyroid hormone-related protein is a gravisensor in lung and bone cell biology. Adv. Space Res. **32**, 1569–1576 (2003)

J.S. Torday, Homeostasis as the mechanism of evolution. Biology (Basel) **4**, 573–590 (2015)

J.S. Torday, W.B. Miller Jr., The resolution of ambiguity as the basis for life: A cellular bridge between Western reductionism and eastern holism. Prog. Biophys. Mol. Biol. **131**, 288–297 (2017)

J.S. Torday, V.K. Rehan, Lung evolution as a cipher for physiology. Physiol. Genomics **38**, 1–6 (2009)

J.S. Torday, V.K. Rehan, *Evolutionary Biology, Cell-Cell Communication and Complex Disease* (Wiley, Hoboken, 2012)

J.S. Torday, V.K. Rehan, *Evolution, the Logic of Biology* (Wiley, Hoboken, 2017)

J. Zhang, P. Hess, A. Kyprianidis, P. Becker, A. Lee, J. Smith, G. Pagano, I.D. Potirniche, A.C. Potter, A. Vishwanath, N.Y. Yao, C. Monroe, Observation of a discrete time crystal. Nature **543**, 217–220 (2017)

W. Zheng, Z. Wang, J.E. Collins, R.M. Andrews, D. Stemple, Z. Gong, Comparative transcriptome analyses indicate molecular homology of zebrafish swimbladder and mammalian lung. PLoS One **6**, e24019 (2011)

Chapter 8
Cybernetics Is a Conversation with the Cosmos

Imagination is more important than knowledge.

Einstein

Introduction

Norbert Wiener was the first to functionally define cybernetics as "the study of control and communication in the animal and the machine." It is largely the basis for our understanding of physiology, as described by Cannon, reduced systematically from the physical to the biochemical, to the cellular-molecular.

von Foerster expanded cybernetics to the second order as awareness of such feedback control (von Foerster, H. (Ed.) (1995). Cybernetics of cybernetics: Or, the control of control and the communication of communication (second ed.). Minneapolis, MN: Future Systems). Others have postulated third- and fourth-order cybernetics, but such speculations are scientifically untestable or refutable. Herein, it is shown that as a manifestation of physiology, cybernetics can be further reduced to cell-cell signaling to maintain homeostasis, bridging Newtonian third-order cybernetics with quantum mechanical fourth-order cybernetics as our title "Conversation with the Cosmos" based on quantum entanglement, constrained by non-localization. If in fact, everything in the cosmos is a product of the Big Bang, including quantum mechanics, then the linkage between third- and fourth-order cybernetics would account for that. And if that is correct, then the sooner we begin operating based on quantum mechanical principles instead of Newtonian mechanics, the sooner we can function effectively based on predictive algorithms, particularly in the realm of artificial intelligence.

For example, evolution based on Darwinian theory is unidirectional, yet time is reversible in quantum mechanics. Therefore, tracking evolution all the way down to quantum mechanics would allow for reverse evolution, as has been demonstrated in *Morphogenesis, Environmental Stress and Reverse Evolution* (2020). This fundamental difference between Darwinian evolution and evolution based on cell-cell communication, seen both synchronically and diachronically, underscores the power of the mechanistic approach, which opens up to the interrelationships

J. S. Torday, *Hormones and Reality*, https://doi.org/10.1007/978-3-030-93691-4_8

between physics and biology, rendering the latter predictive. To quote the Abstract of that article,

> Nowhere are the shortcomings of conventional descriptive biology more evident than in the literature on Quantum Biology. In the on-going effort to apply Quantum Mechanics to evolutionary biology, merging Quantum Mechanics with the fundamentals of evolution as the First Principles of Physiology-namely negentropy, chemiosmosis and homeostasis-offers an authentic opportunity to understand how and why physics constitutes the basic principles of biology. Negentropy and chemiosmosis confer determinism on the unicell, whereas homeostasis constitutes Free Will because it offers a probabilistic range of physiologic set points. Similarly, on this basis several principles of Quantum Mechanics also apply directly to biology. The Pauli Exclusion Principle is both deterministic and probabilistic, whereas non-localization and the Heisenberg Uncertainty Principle are both probabilistic, providing the long-sought after ontologic and causal continuum from physics to biology and evolution as the holistic integration recognized as consciousness for the first time.

Scientific evidence for the connection between the Newtonian and quantum mechanical levels of physiology is limited to the observation regarding the effect of microgravity on parathyroid hormone-related protein signaling (Torday 2003), leading to the notion that it is impossible to achieve due to decoherence. Yet there is evidence that retinal cells can detect singly photons, for example, and Fels has shown that single photons can induce mitosis. And the PIEZO channel can sense pressure, inferring a specific mechanism that could mediate mechanotransduction, like the effect of gravity on lung and bone cells (Torday and Rehan 2003), as well as on yeast (Purevdorj-Gage et al. 2006).

As such, cybernetics can be scientifically tested in toto from the functional to the metaphysical, rendered physical as communication for the first time. This perspective offers common ground for science and art alike, both of which are founded on imagination, striving to transcend Bohm's explicate order, ever-reaching for the implicate, but never quite attaining it because we began as an ambiguity (Torday and Miller 2017) and must remain so to sustain and perpetuate our "Faustian pact."

This perspective on consciousness offers a unifying hypothesis for human endeavor as individuals within social systems, maximizing our evolved traits, particularly imagination and abstract creativity, which distinguish us from all other species. Conventionally, it is assumed that once people mature, imagination diminishes, yet humans are perpetual juveniles as a result of their oversized brains – over the course of human evolution our heads have gotten larger over time, eventually becoming so large that the fetus could not fit through the birth canal; thus, we are born prematurely, with only about 25–30% of adult brain volume at birth in order to compensate. Thus, humans remain in a neotenous juvenile state for the first two to three decades of life in order for the brain to fully develop, if not longer. Consequently, we do tend to remain more imaginative. The accompanying schematic (Fig. 8.1) is a representation of the content of this article.

Cybernetics is basically the mechanism of feedback and as such is a fundamental way of understanding human physiology, transcending the present, going stepwise through the history of the organism, and beyond, to the origin of life itself as the perpetual "Q & A" of life.

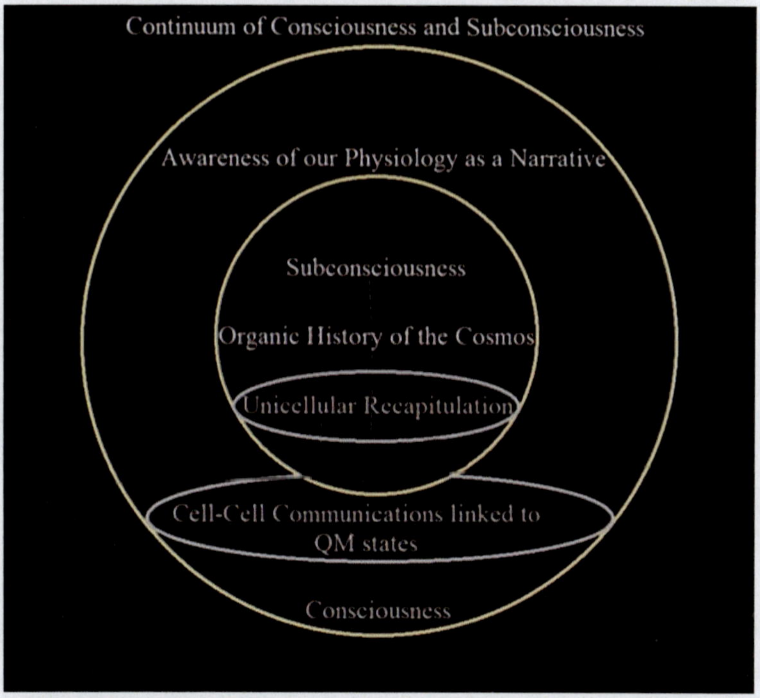

Fig. 8.1 Venn diagram for the content of this article

It has been postulated that the cell-cell signaling mechanisms that have formed multicellular organisms have occurred through symbiogenesis, Lynn Margulis Sagan's theory that when confronted with existential threats over the history of the organism, such factors have been endogenized and assimilated to form our physiology, de facto merging physics and biology. The way in which that occurs can be seen as cybernetic, maintaining or regaining homeostatic control by employing soluble growth factors and their cognate receptors as the mediators of cellular awareness of homeostasis.

But that raises the question as to what the cell is referencing with regard to homeostasis cybernetically. That requires a bit of ingenuity and imagination, given that each step backward over the course of evolution is an exaptation or preadaptation. In this vein, imagination is an exaptation of autopoiesis due to the effect of gravity on planets and cells alike. It is thought that this final common pathway for the autopoietic effect of gravity as one of the laws of nature was produced by the Big Bang. It is the reason why we feel that there is something greater than ourselves.

The question is what preceded the first principles of physiology – negative entropy, chemiosmosis, and homeostasis? It has been proposed that homeostasis is the "equal and opposite reaction" to the Big Bang, based on Newton's third law of motion. Therefore, it is the singularity that preceded the Big Bang that is being referenced, given it is the only thing that existed at that point in the history of the

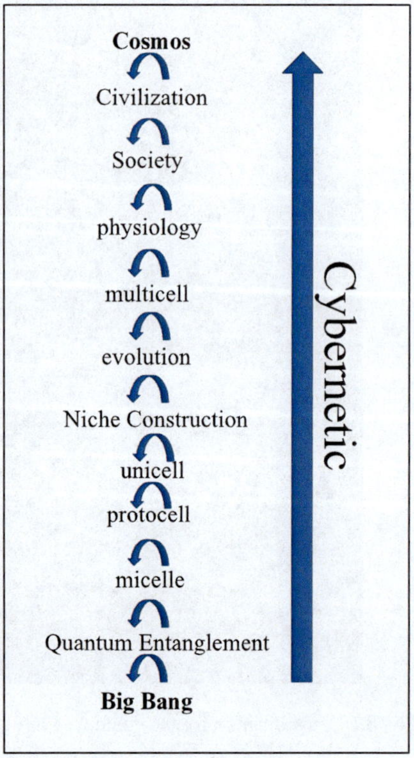

Fig. 8.2 Cybernetics all the way down. Beginning with the Big Bang, cybernetic feedback determines the formation of cells, organisms, social systems, and civilization as a series of "conversations"

cosmos (Hawking 1998). In that vein, the cybernetic conversation between the ever-evolving cell and the cosmos is characterized here as the fourth-order of cybernetics (Fig. 8.2).

The Origin of Life as Our Cybernetic "Dialogue" with the Cosmos

As indicated above, we can trace the cybernetic principle all the way back from everyday Newtonian interactions with our environment to cell-cell signaling as the mechanism connecting quantum mechanical principles of the unicell with the cosmos through quantum entanglement and non-localization. This is how David Bohm characterized our existence as the Explicate and Implicate Orders in *Wholeness and the Implicate Order* (1980) as 'enfolding and unfolding', which are homologous with Symbiogenesis and niche construction.

Now, with the insight that the divide between the self and the cosmos can be seen through the lens of cybernetics, micelles evolved from lipids immersed in water orienting vertically as amphiphiles with negatively and positively charged ends, forming the initial boundary between life and nonlife under the influence of gravity as one of the laws of nature. Gravity, in turn, is produced by planets distorting the fabric of space-time. It has been speculated that the formation of the planets, like that of cells, is also the result of quantum entanglement (Griffiths 2004); thus, planets and cells are homologs, being of the same origin. From that perspective, the origination of life from nonlife due to symbiogenesis could have been predicted based on this perspective, in retrospect, given the premise that evolution is composed of serial exaptations. That raises the question as to where the process of evolution based on endogenizing factors in the environment that posed existential threats emerged from. The accommodation of quantum entanglements referencing nonlocal events in the cosmos is consistent with that process – in retrospect.

The evolution of life in response to changes in the environment is achieved through cell-environment and cell-cell interactions, mediated by soluble growth factors and their cognate receptors, recursively referencing the first principles of physiology – negentropy, chemiosmosis, and homeostasis – based on experimental evidence. Indeed, it is empiricism that has facilitated that continuum, like the Socratic "know thyself," as in Maturana's

poem,***The Student's Prayer***
Don't impose on me what you know,
I want to explore the unknown
And be the source of my own discoveries.
Let the known be my liberation, not my slavery.
The world of your truth can be my limitation;
Your wisdom my negation.
Don't instruct me; let's walk together.
Let my richness begin where yours ends.
Show me so that I can stand
On your shoulders.
Reveal yourself so that I can be
Something different.
You believe that every human being
Can love and create.
I understand, then, your fear
When I ask you to live according to your wisdom.
You will not know who I am
By listening to yourself.
Don't instruct me; let me be.
Your failure is that I be identical to you.

informing us about the true path of learning as self-awareness through cybernetic dialogs, not "top-down" instruction. This is consistent with Gordon Pask's conversation theory (1976) and with the dialogs between Krishnamurti and Bohm.

Knowing yourself can be extended to knowing the world and knowing one's place in the world from a bi-hemispheric viewpoint. Iain McGilchrist has said, "Many 'paradoxes' depend on whether one adopts the left hemisphere (LH) or right hemisphere (RH) way of thinking about, or, more accurately, being in the world." Extremely different types of attention characterize the LH, which decontextualizes and manipulates, while the right hemisphere is complexly interconnected. In this context, knowing yourself implies knowing your hemispheric limitations.

Such a complicated mode of self-awareness is underpinned by our physiologic "history," beginning with our ambiguous origin in negentropy, followed thereafter by the gyrations of the embryo as it goes from zygote to morula, to gastrula, the latter being the stage at which the mesoderm appears between the endoderm and ectoderm, which Wolpert has said is the most important thing we will ever do in our lives. And as deuterostomes, we develop from anus to mouth, in the opposite direction from the earth's force of gravity, which set life as micelles in motion in the first place. Beyond that, there is the process by which we become bipedal, recapitulating the verticalization of lipids immersed in water referred to above, freeing our forelimbs to specialize for flight in the case of birds and toolmaking and language formation in the case of humans.

Both fine motor control and language formation are localized within the area of Broca in the cerebrums of all great apes; humans differ from the other members of the species with respect to bipedalism. Positive selection pressure for an enlarged central nervous system gave rise recursively to a larger and larger brain. Eventually, the conceptus literally "hit a wall" because it could no longer fit through the birth canal, leading to selection pressure for premature birth as hairless apes with only about 30% of adult brain volume. As a result, we humans require 20–30 years postpartum for our brains to reach their total adult volume, explaining our juvenile behavior – instant gratification, risk taking, and narcissism – which lasts for much of our life cycle; it is our social and cultural institutions which have saved us from ourselves through niche construction.

In the context of the "whipsawing" effects of our evolution, Porges's polyvagal theory (Porges 1995) should also be mentioned. It is based on the development and phylogeny of the vagus, the largest of the autonomic nerves, from the adrenal glands forward to the gut, the heart, and ultimately the cranial nerves that enervate the face. It is this phenomenology, along with those other "twists and turns" in our developmental and evolutionary history that may be the foundation for lateralization – left and right brain – acting to accommodate our multifactorial cybernetic orientation to the cosmos.

All of the above is ultimately in service to remaining as proximate to the vectoral force produced by the Big Bang as possible, emanating from the singularity that preceded it. The further we deviate from that relationship, the more "friction" there is between us and our evolutionary trajectory, diminishing the possibility of communing with the cosmos, threatening to break that causal stream of events, which could result in extinction.

This perspective is reminiscent of artists and scientists alike imagining how and why we communicate with the cosmos. Such ideas range from Robert Frost's "I had a lover's quarrel with the world" as a form of communication to John Wheeler's thought that we live in a "participatory universe" (Wheeler 1995) in which observers create the universe (see Fig. 8.3).

It should be noted that these extremes, like virtually every way of thinking about our existence, do not transcend the present view of the world for lack of a systematic way of doing so, whereas the diachronic approach of cell-cell signaling mediated by soluble growth factors and their receptors as the basis for our physiology

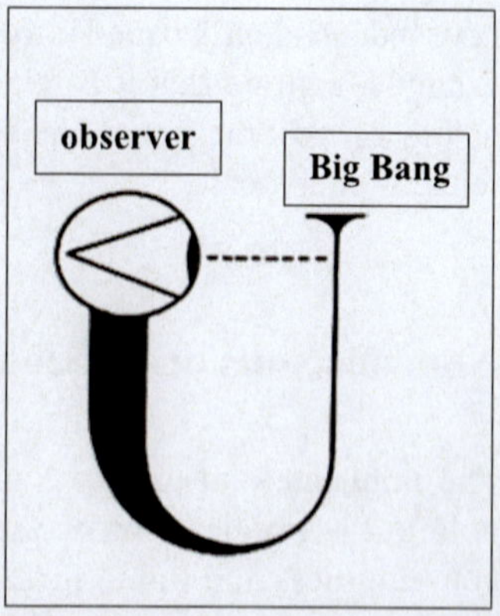

Fig. 8.3 John Wheeler's "participatory universe". Wheeler believed in a universe created by observers

highlighted in this chapter recognizes our evolutionary "history," offering the possibility of transcending space-time materially, seeing the energetic flow of communication as the actual underlying process of life. This perspective is in keeping with Alfred North Whitehead's process philosophy.

In that vein, we should be reminded that the form of the organism is not fixed but is always in flux below the surface, entrained by homeostasis. But when injured or under stress, the organism can morph within or between life cycles based on its physiologic "memory." Waddington relates this in his *The Strategy of the Genes* (1957):

> *It is not enough to see the horse pulling a cart past the window as the good working horse it is today; the picture must also include the minute fertilized egg, the embryo in its mother's womb, and the broken-down old nag it will eventually become.*

This way of seeing life as a flow of energy rather than being composed of material objects is reminiscent of Frescura and Hiley (1984) in their consideration of Bohm's explicate and implicate orders. As particles in quantum mechanics, what is normally thought of as evolution is Bohr's complementarity allowing for the rearrangement of particles through Whitehead process or activity, moving from implicate to explicate order.

Physiology as Ross Ashby's Black Box of Cybernetics

W. Ross Ashby (1956) used the black box metaphor devised by Maxwell to explain cybernetics. The black box contains a mechanism that cannot be seen and is the product of the observer's interaction with the box. Ashby suggested that the black box might not be just a useful device but is universal, suggesting that we never really see what is causing a change, only some explanatory principle we take as a mechanism. This concept of essential obscurity was remarkable for the time and still causes problems for many. In this vein, both Roux (2014) and Cowan (2016) have independently called into question the use of function to define physiology, for example, arguing that it is teleologic and tautologic to do so. Given the black box nature of cybernetics as the fundament of physiology, no wonder its definition remains equivocal.

Nonconscious or Subconscious Cybernetics

The implicate and explicate orders represent boundaries. The micelle punched a hole in the implicate order that generated the explicate order, which allowed for consciousness and life to form from nonconsciousness. Theoretically, a distinction can be made between nonconsciousness and consciousness in relation to first- and second-order cybernetics. Before the micelle came into existence first-order feedback systems were nonconscious. Symbiogenesis sustained the explicate order, and consciousness emerged with life as second-order systems based on awareness. Subconscious (just under conscious) systems are based on the nonconscious systems of the first order. Of course, we are not aware of the many cybernetic relationships, but we know that the third and fourth orders exist based on empiric knowledge of physiology, embryologic development, and phylogenetics, as well as injury-repair, which are all interacting systems. There are always subconscious processes happening in the background of human consciousness, which can become more conscious with awareness of our physiology as a narrative. When Edgar Allan Poe

> says, "All that we see or seem/Is but a dream within a dream," it congers up a sense of the epiphenomenon of the mind derived over generations from prehuman sources.*You are not wrong, who deem*
> *That my days have been a dream;*
> *Yet if hope has flown away*
> *In a night, or in a day,*
> *In a vision, or in none,*
> *Is it therefore the less gone?*
> *All that we see or seem*
> *Is but a dream within a dream.*

(Poe, E. A. *A dream within a dream.* Poetry Foundation. https://www.poetryfoundation.org/poems/52829/a-dream-within-a-dream)

The nonconscious or subconscious levels of cybernetics can come to mind during meditation, or under physiologic stress conditions such as Near-Death Experiences, Out of Body Experiences, Maslow Peak Experiences, or in dreams. As such, those experiences can rise to the level of consciousness, depending upon the circumstances. If the individual is aware of these subconscious or nonconscious levels of cybernetics, they are more likely to incorporate such insights into their conscious being. This is largely due to the mechanistic nature of physiology being mediated by cell-surface receptors that can up- or downregulate to amplify or dampen the signaling, respectively. Therefore, such practices can be positively reinforced personally through education and positively reinforced through social institutions and systems.

Gebser has said, "becoming co-visible in and through man it [the diaphainon] attests to the new mutation by dimensioning in consciousness are integrated and made 'meaningful.' And as these spatio-temporal unfoldings or exfoliations open the possibility for the conscious emergence of the presence of origin, this presence diaphanously enters the realm of conscious and visible effectiveness from its unconscious and invisible state." To understand the presence of the origin, we cannot negate the efficient forms of earlier structures.

On the Cybernetics of Auto-engineering

The relationship of evolution to cybernetics was alluded to above. The mechanism for such adaptive changes is contingent on the force of homeostasis acting to maintain the equipoise of the structure and function involved. Under stress conditions, excessive blood flow through the microcirculation causes the local production of reactive oxygen species (ROS) due to shear stress. Such ROS are known to cause gene mutations and duplications; seen in the context of an evolved trait in compliance with the first principles of physiology, dyshomeostasis will result in failure of cell-cell signaling, followed by the cells involved recapitulating their developmental and/or phylogenetic motifs to reestablish homeostatic control by means of soluble growth factors and receptors recapitulating developmental and phylogenetic motifs, in the spirit of Haeckel's biogenetic law but now with a mechanistic explanation. And in actuality, it also entails Spemann's "organizer," which was a growth factor-receptor mechanism in retrospect. The underlying principle for homeostatic control is cybernetic feedback. The newly established set point is determined by the cellular architecture in what we refer to descriptively as the process of evolutionary adaptation. Of course, the net result would be Darwinian reproductive success, but that is an epiphenomenon.

A classic example of this way of understanding evolution at the cellular-molecular level as cybernetic feedback is recounted in Carroll et al.'s article about the evolution of the glucocorticoid receptor from the mineralocorticoid receptor empirically by cybernetic "trial and error" (Bridgham et al. 2006). Yet the authors characterize it in Darwinian terms as a random gene mutation, whereas it occurred during the water-land transition, an extremely stressful experience existentially linking the hormones for water and electrolyte regulation, i.e., mineralocorticoids, to the hormones for glucose regulation, i.e., glucocorticoids. Elsewhere, it has been recounted that the additional stress of terrestrial life would have increased blood pressure, and in offsetting that, the evolution of the glucocorticoid receptor from the mineralocorticoid receptor would have had a dual advantage of decreasing the mineralocorticoid receptor levels, at the same time stimulating the number of ß-adrenergic receptors, particularly in the pulmonary circulation, allowing for independent regulation of the pulmonary and systemic circulations, which was existential for land adaptation. Of course, this goes against the primary tenet of Darwinian evolution – random mutation. Nonetheless, it is consistent with the duplication of both the glucocorticoid receptor and ß-adrenergic receptor during the water-land transition, begging the question as to whether these were indeed random events or not. If they had been random, they could not recapitulate the process in chronic diseases like emphysema and osteoporosis, for example. The question is further exemplified by the duplication of the parathyroid hormone-related protein (PTHrP) receptor during that very same epoch since PTHrP is necessary for the development of lung alveoli.

This is clearly an example of cybernetic feedback between the organism and its environment, the breeching of land resulting in an increase in the effect of gravity relative to floating in water. Therefore, it is not coincidental that the organ of gas exchange for buoyancy would have experienced increased positive selection pressure, morphing into the lung as the organ of gas exchange for metabolism. This relationship is underscored both by the fact that of the two boney fish phenotypes – physostomous and physoclistous – it is specifically physostomous boney fishes that are our ancestors, having their swim bladder connected to their esophagus by the pneumatic duct, which is homologous with the trachea in land vertebrates. Furthermore, it has been documented that the top 50 transcribed genes in zebrafish swim bladder development are identical with those of the embryonic mouse lung, providing molecular evidence for the homology between the swim bladder and lung (Zheng et al. 2011). Moreover, fibroblast growth factor 10 is necessary for the formation of the smooth muscle sheathing the pneumatic duct in fish and the trachea in mice, further demonstrating the homology between the swim bladder of fish and the lungs of mammals (Korzh et al. 2011).

Discussion

Just to recap, the first order of cybernetics is feedback, the second order is awareness of the feedback, the third order is the cell-cell communications that link cells together to maintain homeostasis, tying the Newtonian mechanics to the quantum mechanical level, and the fourth order is the quantum mechanical communication with the cosmos. Recognition that the ultimate way we commune with the cosmos is at the quantum mechanical scale gains insight to aspects of development and phylogeny untenable otherwise. The fact that soluble growth factors link the cells of multicellular organisms together during the development of the offspring, culminating in integrated physiologic systems now becomes "rational" by realizing that they bridge the quantum mechanical fundament to the Newtonian mechanisms we think of as reality. In point of fact, they do so, but diachronically across space-time as the ties that bind us to the singularity, for that is the ultimate reference point for the unicell as zygote, reconciling the asymmetries caused by the Big Bang. That is the reason why we recursively return to the unicellular state over the course of the life cycle, which has remained dogma up until now. It should be mentioned that physics and chemistry do the same, but their "cycles" are on a cosmic scale, whereas ours are measured in days in the case of mayflies to centuries in the case of giant redwoods.

By reducing consciousness to cybernetics, the processes of inquiry as science or the arts merge together. Scientifically, we formulate a hypothesis as a question, with a binary "yes or no" answer that is testable and refutable. In the arts, a similar process applies, but the methods are much more subjective and indirect, yet the same inquiry into the "truth" hangs in the balance. Even in sports, when you hit a tennis ball to your opponent, you are asking the cosmic question as to whether your opponent knows the "answer" to the shot.

The fundament of life is to query our environment in order to detect novel changes that may or may not affect our survival. Using epigenetic inheritance, we are able to assimilate such subtle changes in the DNA of the egg and sperm through an as yet to be determined mechanism by which they "imagine" similar circumstances in their evolutionary history and incorporate them into our offspring via adducts of DNA. Such chemically bound DNA is subsequently incorporated into the embryo (or not) during conception and subsequent embryonic development. In that way, such "insider knowledge" apprises the organism of on-coming environmental change in order to adapt and better survive our ever-changing surroundings.

The position regarding epigenetic inheritance taken here should be considered in light of the new genetic editing technology CRISPR. Gene mutations may or may not be deleterious, but in either case they are expressed in various places in the body, referred to as pleiotropies. Expressed in one tissue, the mutation may be maladaptive, whereas in another tissue it can be adaptive, so correcting the mutation globally may have unintended consequences.

Perhaps the characterization of the fourth order of cybernetics as a "conversation with the cosmos" is the reason why we universally feel that there is something greater than ourselves. Conventionally, the answer is spiritual, or a sense of the group, as in Martin Buber's *I and Thou*, but perhaps they are derivative of that cybernetic cosmic dialog. An even more universal property would help us feel like we share a common human bond. Lepskiy (2018) has speculated along similar lines about the evolution of cybernetics, but his work is philosophically based, whereas what is being offered herein is based on scientific evidence, allowing it to be objectively testable and refutable.

References

W.R. Ashby. *Introduction to Cybernetics* (Chapman and Hall, London, 1956, 2006)

J.T. Bridgham, S.M. Carroll, J.W. Thornton, Evolution of hormone-receptor complexity by molecular exploitation. Science **312**, 97–101 (2006)

T. Cowan, *Human Heart, Cosmic Heart* (Chelsea Green Publishing, Chelsea, 2016)

F.A.M. Frescura, B.M. Hiley, Algebras, quantum theory and pre-space. Revista Brasileira de Fisica **70**, 49–86 (1984)

D.J. Griffiths, *Introduction to Quantum Mechanics* (Prentice Hall, Hoboken, 2004)

S. Hawking, *A Brief History of Time* (Bantam, New York, 1998)

S. Korzh, C.L. Winata, W. Zheng, S. Yang, A. Yin, P. Ingham, V. Korzh, Z. Gong, The interaction of epithelial Ihha and mesenchymal Fgf10 in zebrafish esophageal and swimbladder development. Dev. Biol. **359**, 262–276 (2011)

V. Lepskiy, Evolution of cybernetics: Philosophical and methodological analysis. Kybernetes **47**, 249–261 (2018)

G. Pask, *Conversation Theory, Applications in Education and Epistemology* (Elsevier, Amsterdam, 1976)

S.W. Porges, Orienting in a defensive world: Mammalian modifications of our evolutionary heritage. A Polyvagal Theory. Psychophysiology **32**, 301–318 (1995)

B. Purevdorj-Gage, K.B. Sheehan, L.E. Hyman, Effects of low-shear modeled microgravity on cell function, gene expression, and phenotype in Saccharomyces cerevisiae. Appl. Environ. Microbiol. **72**, 4569–4575 (2006)

E. Roux, The concept of function in modern physiology. J. Physiol. **592**, 2245–2249 (2014)

J.S. Torday, Parathyroid hormone-related protein is a gravisensor in lung and bone cell biology. Adv. Space Res. **32**, 1569–1576 (2003)

J.S. Torday, W.B. Miller Jr., The resolution of ambiguity as the basis for life: A cellular bridge between Western reductionism and eastern holism. Prog. Biophys. Mol. Biol. **131**, 288–297 (2017)

J.S. Torday, V.K. Rehan, Mechanotransduction determines the structure and function of lung and bone: A theoretical model for the pathophysiology of chronic disease. Cell Biochem. Biophys. **37**, 235–246 (2003)

J.A. Wheeler, *At Home in the Universe* (Springer, New York, 1995)

W. Zheng, Z. Wang, J.E. Collins, R.M. Andrews, D. Stemple, Z. Gong, Comparative transcriptome analyses indicate molecular homology of zebrafish swimbladder and mammalian lung. PLoS One **6**, e24019 (2011)

Chapter 9
Atavisms Redux

Introduction

Brian Hall has written extensively about atavistic traits, like limbs in whales and toes in horses (Hall 1995). These are the superficial phenotypic traits that can be seen with the naked eye, whereas there are many internal atavisms at the cellular-molecular level, such as the watershed realization that the fish swim bladder is the precursor to the lung, as a homology, not the gill, which is an analogy. Or that bottleneck encountered by evolutionists trying to trace the evolution of the central nervous system in vertebrates, seeing no connection to invertebrates. That is until Nick Holland published his "skin-brain" hypothesis (Holland 2003), pointing out that invertebrates have their central nervous systems in their skin. That realization allowed for the continuum from the cell membranes of unicellular organisms to the highly evolved brains of humans and everything in between.

Let Us Begin with the Unicell

Key to seeing these deep, historic atavistic homologies is to begin with the unicell, looking in the forward direction, multicellular traits having evolved through cell-cell signaling mechanisms that aid in "seeing" how such traits have emerged by "imitation." It is essential that we begin seeing our origin and fate in their true direction in order to better comply with the evolutionary mandate and purpose of our life cycle – to scout out novelty in our environment in order to provide advanced, meaningful notice to our progeny or what is referred to as epigenetic inheritance. It is that process that informs us of our purpose, so it will be enumerated. Following birth, the offspring begins interacting with its environment, on the breast, crawling, toddling, as the brain grows from its initial state at only 25% of its total volume. Seen

J. S. Torday, *Hormones and Reality*, https://doi.org/10.1007/978-3-030-93691-4_9

descriptively, Piaget thought that we had to have this long period of childhood development in order to accommodate the growth and differentiation of our large brains, failing to recognize the underlying interactive, epigenetic process – as the developing child interacts with its environment, it is assimilating what it experiences in its proper context, which is then absorbed through the circulation if it is a physical entity, transported to the egg or sperm. During the process of meiosis – going from 46 chromosomes to 23 – there is a mechanism by which the cells deliberate whether some epigenetic "mark" is significant to its physiologic well-being or not, undoubtedly ultimately based on its genetics as the "truth" of its evolutionary history. Once that process is completed the relevant DNA code(s) is biochemically reacted with methyl, ubiquitin, or other chemicals to modify the "readout" of the genetic information involved. As a result, the offspring will phenotypically change as a function of the genetic modification, in turn altering its interactions with its environment in order to effectively adapt to the presaged environment.

Internal Selection

The internal changes that have occurred over the course of vertebrate evolution are revealed by their relationships to environmental changes. The most dramatic adaptive changes occurred during the water-to-land transition, caused by the "greenhouse effect" that partially dried up the ocean, as alluded to above. There were three specific gene duplications that occurred during that period – the parathyroid hormone-related protein receptor (PTHrPR), the glucocorticoid receptor (GR), and the β-adrenergic receptor (βAR). All three of these receptor duplications were essential for successful adaptation to land, in retrospect. The interrelationships between these receptors have been unmasked by deletion of the PTHrPR, revealing their allostatic "history." Knockout of the PTHrPR causes several structural-functional deficiencies, the most prominent being the failure to form alveoli in newborn mice, resulting in their death shortly after birth (Rubin et al. 2004). In conjunction with the lung phenotype, the kidneys fail to form fully functional glomeruli, the skin remains immature, and the skeleton fails to calcify, consistent with their fish origins. All four of these physiologic traits were existential for the ability of fish to adapt to land. The specificity of this process is underscored by the fact that it was specifically the physostomous boney fish that gave rise to land vertebrates, i.e., those boney fish that have a pneumatic duct or "tube" connecting the esophagus to the swim bladder. The pneumatic duct is the homologue (of the same origin) of the trachea, and it facilitates the inflation and deflation of the swim bladder, allowing the fish to efficiently adapt using buoyancy in its watery environment. Its cousins the physoclistous boney fish inflate and deflate their swim bladders by direct exchange of gases with the circulation, which is distinctly different from physostomous fish.

Being forced out of water onto land due to the depletion of oxygen in the water, several physiologic changes had to occur – the organism had to have a skeleton that could support it under the increased effect of gravity on land; it had to be able to breathe air; the integument had to be able to retain water and electrolytes more efficiently. All of the physiologic changes caused by increased PTHrPR signaling resulting from the doubling of the number of such receptors amplified those adaptations.

Evolution of the Glucocorticoid Receptor

As for the evolution of the GR during the water-to-land transition, it has been shown to have been derived from the mineralocorticoid receptor (MR) as a result of the addition of three amino acids to the binding site of the MR (Bridgham et al. 2006). How and why that occurred during the water-to-land transition will become apparent when taking into consideration the context. Under the physiologic stress of life on land due to an increased gravitational effect, blood pressure would have increased spontaneously. To alleviate that stress on the circulatory system, it would have been advantageous to reduce MR signaling, which increases blood pressure. Through "trial and error," the modification of the MR binding site by addition of amino acids gave rise to the GR.

Evolution of the βAR

As for the βAR, its duplication gave rise to independent regulation of the lung circulation from the systemic circulation. That was critically important for the successful evolution of the lung from the swim bladder, given that the lung must be able to increase oxygenation "on-demand." The enhanced GR signaling acted synergistically by increasing the number of βARs.

Suffice it to say that none of these phenotypic changes that were necessitated by the water-to-land transition by vertebrates have been applied to the evolution of Tiktaalik, both because there is no "fossil" record of such changes and because Darwinian evolution is strictly due to random gene mutations (Huxley 1942), whereas the functional genomic interrelationships enumerated above are causal. This difference in perspectives is due to the two prevailing views of evolution, from its results in hindsight to its stepwise changes from unicellular to multicellular. The former would naturally yield binary relationships in hindsight, whereas the latter would lend itself to seeing the process prospectively as multiple "options" for adaptation, being resolved by the prevailing conditions as a continuum. The latter was achieved largely by merging developmental and phylogenetic data as one based on their shared origins in cell-cell signaling mechanisms.

Pleiotropy

In conventional, descriptive biology, the physiologic interrelationships cited above are recognized as pleiotropies and thought of anecdotally. Conversely, with the advent of molecular biology, such pleiotropic interrelationships can now be seen as causal once the constraint of Darwinian evolution as random genetic mutations is shed in favor of a mechanistic cellular approach. In reality, all of contemporary biology and medicine is seen through the lens of cell biology, but the underpinnings as evolved traits unfortunately remain entrenched in description instead of mechanism. It is unfortunate because the latter reveals biologic interrelationships untenable through after the fact reasoning and hampers the effective diagnosis and treatment of disease as symptoms that must be eradicated rather than identifying the etiology in order to effectively treat diseases.

The Role of Bipedalism

Bipedalism refers all the way back to the vertical orientation of lipids in water, perhaps being exapted due to the increase in body temperature, referencing the sun warming the lipids in the air-water interface primordium. So, for example, neuregulin mediates lipid adaptation in the alveoli (Dammann et al. 2003) and myelinization of neurons (Birchmeier and Bennett 2016), reflecting the positive selection pressure for both in bipedal adaptation and the central nervous system. The lipid changes in the composition of lung surfactant attest to such interrelationships, becoming more effective in surface tension reduction at 41 °C in warm-blooded organisms (Suri et al. 2013). The causal relationship is demonstrated by an experiment with map turtles, which adapt their surfactant to the temperature of the atmosphere. Briefly, the lipid composition of the lung surfactant changes in accord with the change in ambient temperature, optimizing lung alveolar function (Lau and Keough 1981).

Heart-Hand Relationship: *Ciona Intestinalis*

A good example of what might be thought of as an atavistic trait repurposed for evolution is the heart, seen from the perspective of the development of *Ciona intestinalis* (Davidson and Levine 2003). The stem cells for the heart derive from the tail, linking locomotion with circulation, "beating" of the heart coming from the back and forth motion of the tail. And locomotion refers all the way back to the introduction of cholesterol into the cell membrane, increasing its fluidity, in turn increasing cytoplasmic flow, promoting locomotion. And locomotion refers all the way forward to bipedalism in humans but again referencing the perpendicular orientation of lipids in water.

Goodpasture Syndrome

The Goodpasture syndrome isoform of type IV collagen, alpha3(IV)NC1, is not expressed in worms or fish but is expressed in frogs, chickens, mice, and humans (MacDonald et al. 2006). Functionally, the alpha3(IV)NC1 type IV collagen isotype has greater capacity for inhibiting leakage of fluids and electrolytes through the skin, lung alveoli, and kidneys, so it is adaptive for transitioning from water to land, but there can be pathologic consequences. Some people develop an autoimmune reaction to the Goodpasture isoform of type IV collagen, which can cause illness and even death in some cases.

At any rate, the abovementioned evolved traits can be seen as atavisms in descriptive biologic terms but are actually the way that cellular evolution works. The advantage of the latter perspective is that such so-called atavisms are no longer seen as anecdotal. They are part of the continuum of internal selection for homeostasis. As such, the linkages between them can be understood as the underpinnings of physiologic evolution.

Alignment of Hox Genes

In most bilaterians, Hox genes are expressed in staggered domains along the head-to-tail axis of the embryo (Gaunt 2018), suggesting that their role in specifying position is a shared, ancient feature.

References

C. Birchmeier, D.L. Bennett, Neuregulin/ErbB signaling in developmental myelin formation and nerve repair. Curr. Top. Dev. Biol. **116**, 45–64 (2016)

J.T. Bridgham, S.M. Carroll, J.W. Thornton, Evolution of hormone-receptor complexity by molecular exploitation. Science **312**, 97–101 (2006)

C.E. Dammann, H.C. Nielsen, K.L. Carraway 3rd, Role of neuregulin-1 beta in the developing lung. Am. J. Respir. Crit. Care Med. **167**, 1711–1716 (2003)

B. Davidson, M. Levine, Evolutionary origins of the vertebrate heart: Specification of the cardiac lineage in Ciona intestinalis. Proc. Natl. Acad. Sci. U. S. A. **100**, 11469–11473 (2003)

S.J. Gaunt, Hox cluster genes and collinearities throughout the tree of animal life. Int. J. Dev. Biol. **62**, 673–683 (2018)

B.K. Hall, Atavisms and atavistic mutations. Nat. Genet. **10**, 126–127 (1995)

N.D. Holland, Early central nervous system evolution: An era of skin brains? Nat. Rev. Neurosci. **4**, 617–627 (2003)

J. Huxley, *Evolution: The Modern Synthesis* (Allen & Unwin, Melbourne, 1942)
M.J. Lau, K.M. Keough, Lipid composition of lung and lung lavage fluid from map turtles (Malaclemys geographica) maintained at different environmental temperatures. Can. J. Biochem. **59**, 208–219 (1981)
B.A. MacDonald, M. Sund, M.A. Grant, K.L. Pfaff, K. Holthaus, L.I. Zon, R. Kalluri, Zebrafish to humans: Evolution of the alpha3-chain of type IV collagen and emergence of the autoimmune epitopes associated with Goodpasture syndrome. Blood **107**, 1908–1915 (2006)
L.P. Rubin, C.S. Kovacs, M.E. De Paepe, S.W. Tsai, J.S. Torday, H.M. Kronenberg, Arrested pulmonary alveolar cytodifferentiation and defective surfactant synthesis in mice missing the gene for parathyroid hormone-related protein. Dev. Dyn. **230**, 278–289 (2004)
L.N. Suri, A. Cruz, R.A. Veldhuizen, J.F. Staples, F. Possmayer, S. Orgeig, J. Perez-Gil, Adaptations to hibernation in lung surfactant composition of 13-lined ground squirrels influence surfactant lipid phase segregation properties. Biochim. Biophys. Acta **1828**, 1707–1714 (2013)

Chapter 10
We Are All Citizens of Gaia

We are riding on a railroad singing someone else's song.

James Taylor

Gaia Is US

James Lovelock expressed the idea that the earth is an organic whole in his groundbreaking book *Gaia Theory* (Lovelock 1972), which we all sense is true as its citizens, but that is only a gut feeling, a conjecture. However, the emerging concept of niche construction scientifically interconnects the environment with the earth's flora and fauna mechanistically, stepwise, offering the opportunity to understand our relationship to the cosmos rationally.

Nominally, niche construction is the process by which organisms form their own personalized environments, a phenomenon which was first documented by Darwin himself (1881), observing that earthworms retain their water-adapted kidneys, or nephridia, on land by actively modifying the soil around themselves both physically and chemically. Other organisms like beavers build dams, various animals build burrows, and humans build homes, towns, cities, states, and nations, or as Simon Conway Morris has characterized it, "First there were bacteria, now there is New York."

And Nicholas Christakis has quantified human niche construction, formulating his social contagion theory (Christakis and Fowler 2013), quantifying interconnections between human communities, as described using mathematical modeling.

But the underlying mechanism for such "contagion" communities remains unknown. One potential explanation is that the unicell itself was the first niche construction (Torday 2016), based on the endosymbiosis theory promulgated by Lynn Margulis to great acclaim. According to that theory, when unicells were confronted with an existential threat to their environment, they coped by endogenizing it, compartmentalizing it, and thus assimilating it as an add-on to their evolving physiology. In this way, the internal cellular "environment," Claude Bernard's *milieu interieur*, was generated by physical elements that must comply with the laws of nature in the same way that they do in the external environment, forming a literal functional continuum from physics to physiology.

J. S. Torday, *Hormones and Reality*, https://doi.org/10.1007/978-3-030-93691-4_10

In that vein, the atom and unicell are homologues, meaning that they are of the same origin from the singularity that preceded the Big Bang. Both exhibit deterministic and probabilistic characteristics. For example, the hydrogen atom is comprised of an electron in electrophysical balance with a proton; in the case of the unicell, it is comprised of negative entropy and chemiosmosis, which are determined, and homeostatic control, which is probabilistic. The atom "evolved" from the Big Bang, which subsequently formed the template for the unicell.

As for the scalable property of niche construction, it derives from the concept of the phenotype as agent (Torday and Miller 2016). Briefly, over the course of their life cycles, organisms collect epigenetic "marks" from their environment, informing the next-generation offspring of changes by modifying the DNA of the egg and sperm of the parents. Thus, the offspring become agents for repeating this process generation upon generation as an effective way to adapt.

Darwinian evolution is predicated on random mutations and natural selection for fitness. However, based on the inheritance of so-called epigenetic "marks" obtained directly from the environment that inform the germ cells by changing their DNA "readout" in order for the offspring to adapt, the phenotype behaves actively as an agent[4], monitoring for changes in the environment and informing the organism's germ cells (egg, sperm) and ultimately the offspring. It is this active "wandering" behavior of the phenotype, described in great detail by Maura O'Conner in her book *Wayfinding* (2019), that generates structurally and functionally integrated ecologic niches composed of one and the same stuff.

As such, niche constructions grow into progressively larger and larger interconnected communities, ultimately blanketing the earth, forming what Lovelock has referred to as Gaia. This intimate, mechanistic interrelationship between the biosphere and the physical environment is the basis for our innate familiarity with and love for Mother Earth.

Phenotypic Variation as Agency for Epigenetic Inheritance

Niche construction promotes evolution by creating conditions for integrated epigenetic interactions. It is an attractive alternative to the "chance" nature of Darwinian evolution. By fostering phenotypic variation, epigenetics forms the environmental niches in which the organism can thrive among and with its niche mates. Characterizing the phenotype as an "agent" for the acquisition of epigenetic marks, the role of the organism is very different from how Darwin's depiction of evolution. Phenotypic agency is in service to the effective integration of the organism with its environment in order to continuously monitor for external changes; in the case of Darwinian selection, random mutations modify the adult based on reproductive success. By focusing on the adult stage of the life cycle, the evolutionary significance of the developmental phase during which meiosis assimilates the epigenetic "marks" is framed but not explained, leaving us still adrift. Yet it is through the underlying cell-cell signaling mechanism that Haeckel's "ontogeny recapitulates phylogeny" has been realized.

Gaia theory envisions the earth as a self-regulating complex system composed of the biosphere, the atmosphere, the hydrosphere, and the pedosphere, tightly coupled together in a perpetually evolving system. As a holistic entity, Gaia seeks a physical and chemical environment optimized and perpetuated for the sake of life.

Gaia has evolved via cybernetic feedback between the organic and inorganic, leading to stable homeostatic conditions for habitability. The surface conditions of the earth have become optimal for life, resulting from interactions between life forms, particularly microorganisms and inorganic elements. These dynamic homeostatic interactions have established a global control system for the earth's surface temperature, atmosphere, and ocean salinity, driven by the thermodynamic disequilibrium of the earth system.

The earth's homeostasis, generated by living forms, had already been recognized in the field of biogeochemistry and is being studied in other scientific fields such as earth system science. The Gaia hypothesis stands out because it is based on the precept that homeostatic equilibrium is an active mechanism for maintaining and perpetuating the optimal conditions for life.

Humanistically, Gaia hypothesis has the potential for allowing us to appreciate our biologic origins and history as an integral, holistic whole. James Lovelock referred to this as an attitude adjustment that beseeches our collective conscience. Such a basic insight to us as a species among species, literally formed by the environment is a long-awaited change in our perception, which would divorce us from our belief in the anthropic principle that we are "in" this environment and not "of" it.

Gaia theory is bolstered by other processes by which the biota are integrated with the earth. Stigmergies, for example, are detectable traces of activities in the environment that stimulate subsequent actions as a way of understanding collective processes such as social networks.

Stephen J. Gould was famous for asking if we could replay the evolutionary tape, but if we have evolved from an ever-changing environment, the answer would unequivocally be "no." Yet we have previously survived a "greenhouse effect," so if we are destined to endure that process again due to climate change, our phenotype would have to change or we would probably become extinct. Even inhabiting another "earth" would not solve the problem if we do not recognize how and why we have come to this place and time as humans.

On the Evolution of Metazoans

Unicellular organisms dominated the earth for 3.5 billion years before multicellular organisms evolved. In all likelihood, this transition was due to the challenge presented by bacteria as pseudo-multicellular organisms through the formation of biofilm and quorum sensing, the capacity of bacteria to communicate with one another for higher order function. These practices posed a threat to unicellular eukaryotes, our forebears, that was met by eukaryotes in turn forming multicellular organisms using cell-cell signaling for metabolic cooperativity, leading to the flourishing of metazoans over the course of the last 500 million years.

Consciousness as the Product of Gaia: Why We Inherently Care About Mother Earth

The fundamental interrelationship between the internal cytoplasm of the cell and external physical environment derives from its common origin in the aftermath of the Big Bang, both the inanimate and the animate having to comply with the laws of nature. The physical and chemical dualities produced by the Big Bang resolved themselves through reactions controlled by homeostasis, hence the "equals" sign between the energy and mass of the reactants on the left and the energy and mass of the products on the right (see Fig. 10.1). Homeostasis is the residual of the "equal and opposite reaction" to the Big Bang, based on Newton's third law of motion. If it were not for that "equal and opposite reaction to the Big Bang," there would literally be no matter in the cosmos, only chaotic free energy.

Homeostasis governs physical and chemical reactions, as well as the biologic reactions formed by cell-cell signaling that evolved from them through endosymbiogenesis, allowing for the equivalency of reactants and products. The signaling mechanisms involved are mediated by both high energy phosphate and inositide cascades, representing biologically based chemical reactions. Parenthetically, Alfred North Whitehead's "process philosophy" focuses on the interrelationships between energy states, considering matter to be a transient by-product. Thus, the dualities produced by the Big Bang are resolved, culminating in Gaia.

And in the context of consciousness and Gaia, it should be borne in mind that the unicell is conscious according to Arthur Reber's *The First Minds* (2018). Not to mention plants, which are also conscious in the sense that they are aware of both their surroundings and themselves, as elegantly expressed by Trewavas et al. (2020). This idea may seem strange and foreign at first, but consider the fact that a paramecium reacts to a stimulus with a calcium burst in the same way that a neuron does physiochemically, so why not consider that these phenomena are basically one and the same? Based on Occam's razor, the simplest answer is usually correct.

This way of thinking about cells and consciousness connecting Gaia and the cosmos is counterintuitive, yet there is justification in thinking in such out-of-the-box terms. For example, Libet's experiment showing that there is a time delay between reacting to a stimulus and that event registering in the brain implies that there is not a one-to-one relationship between stimulus and response between our

$$\mathbf{E_1 + M_1 = E_2 + M_2}$$

Fig. 10.1 Balanced reaction. Energy (E) and mass (M) on either side of the equals sign is balanced by homeostasis environment and our brain or Mashour's observation that when patients recover from general anesthesia, they recapitulate the evolution of the brain from amphibians to mammals, opening up to the cellular path for such a process. And experiments demonstrating the direct effects of microgravity on cellular structure and function speak to the deep cosmologic relationship between life and the physical world.

The effect of microgravity on yeast is particularly compelling in this light since they lose their capacity to relate to the environment because they cannot polarize, i.e., generate a directed calcium wave or reproduce, which is the hallmark of life. All of the above justifies consideration of what consciousness actually constitutes as a transcendent awareness across space-time, not merely being conscious of our mundane surroundings and self in the here and now.

Morality as Gaia

In retrospect, we have evolved through "cellular cooperativity," yet Darwinian evolution focuses on competition as the driving force behind evolution. In a biography of Derek Parfit in *The New Yorker* magazine, entitled "How to be Good" (Marquhar 2011), the British philosopher and ethicist pauses to express his confusion regarding the paradox of Darwinian "survival of the fittest" with being good. That is because Darwinian evolution is based on descriptive biology, seeing evolution as a consequence of the process of reproduction, when in reality "descent with modification" is the long-term result of life accommodating the laws of physics through cell-to-cell signaling for Gaia niche construction (Torday 2016). The juncture at which protozoans and metazoans "split," referred to as metabolic cooperativity, marked a "sea change" founded on the golden rule, "do onto others," not "kill or be killed." The former is the embodiment of cell-cell cooperation for metabolic drive over the course of evolution, whereas the latter is based on superficial analogical comparisons with animal breeding popular in Victorian England during Darwin's lifetime.

Climate Change, Gaia, and Us

Up until now, the interrelationships between matter and life on earth have been governed by natural law. However, Man's ingenuity has disrupted that harmonious relationships, causing ever-greater upheavals in what is now being touted as the Anthropocene. In essence, we have moved from a natural progression of homeostatic principles that have scalarly interconnected matter and man to an artificial system of causation disconnected from our origins. This course of action violates the singularity of nature (Torday 2019), throwing us into a relationship with ourselves and our surroundings that is man-made, without any bearing on our beginnings or evolutionary arc. In other words, we have taken command of our fate, independent of the mechanisms that facilitated our evolution that have gotten us to this point in human biologic history. As a result, we are now on our own, up a proverbial creek without a paddle, still driven by our survivalist narcissism.

Coda

The epigraph

at the beginning of this article is the opening lyrics of a James Taylor song, entitled.**Riding on a Railroad**

We are riding on a railroad, singing someone else's song
Forever standing by the cross road.
Take a side and step along.
We are sailing away on a river to the sea.
Maybe you an' me can meet again.
We are riding on a railroad, singing someone else's song
Sing along.
Time to time I tire of the life that I've been leading
Town to town, day by day
There's a man up here who claims to have his hands upon the reins.
There are chains upon his hands and he's riding upon a train.
We are riding on a railroad, singing some else's song
Forever standing by the cross road.
Take a side and step along.
We are sailing away on a river to the sea.
Maybe you an' me can meet again.
We are riding on a railroad, singing someone else's song.

The song is an anthem for our communal cognitive dissonance as citizens of Gaia, essentially practicing self-deception in lieu of the facts. In that spirit, this article is intended to make us aware of the fundamental, intimate interrelationship between physical matter and life that acts as our enfranchisement as citizens of the earth, metaphorically expressed as Gaia theory. No longer should Gaia merely be just a spiritual thought; it is literally "in our DNA."

The intimate relationship between consciousness and cosmology is perhaps best expressed by a story that David Foster Wallace used in his commencement address to the 2005 graduating class of Kenyon College: "There are these two young fish swimming along, and they happen to meet an older fish swimming toward them who nods at them and says, 'Morning, boys. How's the water?' And the two young fish swim on for a bit, and eventually one of them looks over at the other and asks, 'What the hell is water?'"

That story epitomizes why we have formally been debating the question of consciousness since the ancient Greeks but still have not come to closure. Our ongoing confusion about what consciousness is exemplified by Andy Clark's "disembodied mind" and David Chalmer's "hard question" regarding the qualia we see as "red" when we feel severe pain. Even in Jared Diamond's book *Collapse* (2011), documenting the successes and failures of civilizations that have or have not lived in harmony with their environments, respectfully, the take-home message is only by example, not principle.

The resolution expressed in this essay is based on an empiric evolutionary approach to physiology, the aggregate being what we think of as consciousness. Once the cellular mechanistic interrelationship between biology and quantum mechanics (Torday 2018) as matter, mind, and consciousness is realized, the connection between man and cosmos becomes evident. And if we were to understand and appreciate this view of reality, perhaps we would be better stewards of our environment because our very lives as human beings depend upon it. To do otherwise would mean divesting ourselves of our birthright.

Up until now we have unconsciously complied with the dictates of evolution, but because of the emergence of bipedalism, the wheel, moveable type, the World Wide Web, artificial intelligence, and artificial genetics, we are on the brink of deviating from our arc of evolution, set in motion by the Big Bang, referencing the singularity of nature (Torday 2019) that preceded it. The irony is that that realization was made feasible by the invention of the scientific method as "the only way we know what we do not know." Perhaps the realization of the intimate relationship between matter, man, and morality expressed herein will act as a "wake-up call" for reason to prevail over our tendency to do things that undermine our very existence.

Evolution is a grand experiment of interactions between us and the ever-changing environment. Whether it succeeds or fails is dependent on whether we adhere to its principles or not. The Biologic imperative is usually defined as food, shelter, and reproduction, based on the superficialities of our observation of our surroundings, which David Bohm referred to as the explicate order in his book *Wholeness and the Implicate Order* (1980). It is only through experimentation that we can transcend that subjective explicate perspective and move ever- closer to the implicate order that lies beyond it.

References

D. Bohm, *Wholeness and the Implicate Order* (Routledge and Kegan Paul, London, 1980)

N.A. Christakis, J.H. Fowler, Social contagion theory: Examining dynamic social networks and human behavior. Stat. Med. **32**, 556–577 (2013)

C.R. Darwin, *The Formation of Vegetable Mould, Through the Action of Worms, with Observations on Their Habits* (John Murray, London, 1881)

J.E. Lovelock, Gaia as seen through the atmosphere. Atmos. Environ. **6**, 579–580 (1972)

L. Marquhar, *How to Be Good* (The New Yorker, New York, 2011)

M. O'Connor, *Wayfinding* (Affirm Press, Melbourne, 2019)

A. Reber, *The First Minds* (Oxford University Press, Oxford, 2018)

J.S. Torday, The cell as the first niche construction. Biology (Basel) **5**(2), 19 (2016)

J.S. Torday, Quantum mechanics predicts evolutionary biology. Prog. Biophys. Mol. Biol. **135**, 11–15 (2018)

J.S. Torday, The singularity of nature. Prog. Biophys. Mol. Biol. **142**, 23–31 (2019)

J.S. Torday, W.B. Miller, Phenotype as agent for epigenetic inheritance. Biology (Basel) **5**(3), 30 (2016)

A. Trewavas, F. Baluška, S. Mancuso, P. Calvo, Consciousness facilitates plant behavior. Trends Plant Sci. **25**, 216–217 (2020)

Chapter 11
The Universal Biologic Basis for Moral Behavior: Personal and Societal Alike

First food, then morals.
Bertolt Brecht, in The Three Penny Opera
A subtle chain of countless rings
The next unto the farthest brings;
The eye reads omens where it goes,
And speaks all languages the rose;
And, striving to be man, the worm
Mounts through all the spires of form.

Ralph Waldo Emerson, in Nature

Introduction

Derek Parfit, the British philosopher, was featured in a biographical profile in *The New Yorker* magazine in 2011 (MacFarquhar). The title of the article was "How to be Good," and in it Parfit expressed his frustration in trying to reconcile Darwinism and empathy – how do you ascribe to "survival of the fittest" and behave decently at the same time? This is one of the multitude of dichotomies that have formed from the ambiguous origins of life, compensated for by deceptions, both deception of self and deception of others, which have caused confusion like that expressed by Parfit. Such paradoxes are due to our lack of understanding of our origins and the ways we have devised to cope with those ambiguous beginnings. In his *The Better Angels of Our Nature*, Steven Pinker (2012, Penguin, New York) addresses this by ascribing it to the difference between our innate understanding of "right and wrong" and what we are actually confronted with due to individual interpretations of morality.

If we frame the question of morality in biologic terms, beginning at the origin of life, perhaps the interrelationships between biology, physics, and morality are more substantive. Protocells evolved from lipids immersed in water, delivered to the earth's surface as frozen snowball-like asteroids. Lipid amphiphiles spontaneously form micelles in water or primitive cells, principally able to form an internal environment (Fig. 11.1). The generation of energy within the cell occurred spontaneously, due to chemiosmosis – referencing the quantum entanglement that refers to nonlocality in the Cosmos – which is the capacity of internal semipermeable membranes to partition negatively and positively charged ions on either side of themselves, creating a flow of electrons. Such bioenergy was necessary to maintain the

J. S. Torday, *Hormones and Reality*, https://doi.org/10.1007/978-3-030-93691-4_11

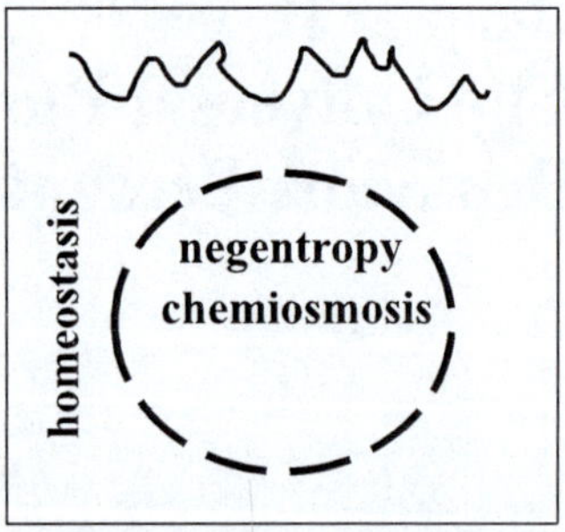

Fig. 11.1 Formation of micelles. Snowball-like asteroids delivered water and lipids to the surface of the earth. Lipids immersed in water form micelles, which provided negentropy, supported by chemiosmosis, and controlled by homeostasis

negative entropy within the cell, or negentropy, regulated by homeostasis, the dynamic mechanism for maintaining energetic/negentropic balance.

Evolution of multicellular organisms required a way of coordinating homeostasis across tissues, organs, organ systems, and overall physiology (Torday and Rehan 2017), accomplished through endocrine and neuroendocrine mechanisms of allostasis. Holistic integration of these processes can be seen through the seminal role of cholesterol, initially as the means of thwarting the effect of rising levels of oxygen in the atmosphere, lipids acting as antioxidants, and then through positive selection pressure, for cholesterol as the parent compound for the steroid hormones such as glucocorticoids, mineralocorticoids, testosterone, and estrogen as well as cholecalciferol or vitamin D3.

"Physiology as Our Selves"

Unicellular organisms dominated the earth for the first 3.5 billion years. It was only 600 million years ago that unicellular eukaryotic organisms began mimicking prokaryotic pseudo-multicellularity in the form of biofilm and quorum sensing that unicellular eukaryotes evolved true mechanisms of cell-cell communication in order to survive in such a competitive environment. Cell-cell signaling mechanisms gave rise to multicellular organisms embryologically and phylogenetically, culminating in physiologic systems for metabolism, respiration, locomotion, and consciousness. For the sake of parsimony, it behooves us to consider the possibility that it also gave rise to moral behavior as the basis for both man's law and God's law alike.

Consciousness as the Endogenization of the External Environment and the Laws of Nature

The underlying principle behind physiologic evolution is the endogenization of the external environment, replete with the laws of nature – gravity, electrochemistry, and thermodynamics. The vertical integration of such physiologic principles based on natural law could have lain the groundwork for moral behavior as the embodiment of nature's laws.

But as expressed elsewhere, the origins of life are ambiguous due to the circumvention of the second law of thermodynamics. Life has coped with that unlawful origin through deception or cheating as the means of survival. However, once that interrelationship was recognized, it offered the opportunity for self-realization regarding the misapprehension of the fundament of life, allowing the opportunity to function out of knowledge instead of ignorance and fear.

Homeostasis as the Fundament of Morality

The homeostatic force originated as the "equal and opposite reaction" to the Big Bang, i.e., Newton's third law of motion. Without that force, there would be no matter in the cosmos, only free energy. Fast forward to the initiation of life, founded on the first principles of physiology (referred to above), the first two principles – negentropy and chemiosmosis – being deterministic, homeostasis being probabilistic, provide free will. It is the latter that offers the opportunity for moral behavior in service to the first two principles as constraints on the organism. As stated elsewhere, this interrelationship is homologous with the Pauli exclusion principle, which dictates the spin state of electrons within an atom. There are four quantum numbers, the first three being deterministic, the fourth being probabilistic. It is hypothesized that since life evolved from matter, it ascribes to such quantum mechanical principles. Similarly, life began as an ambiguity of entropy within and outside of the cell (Torday and Miller 2017), which is homologous with Heisenberg uncertainty principle. All that to say that moral behavior is based on such fundamental physical principles, offering the opportunity for life to exist in the cosmos.

Morality as an Atavistic Trait of the Unicell

Hobson et al. (2014) have described the phenomenon of "brain cooling" during rapid eye movement sleep. It has been speculated that this is the way that higher consciousness formed in service to the evolutionary strategy. The relationship between control of body heat, memory and evolution references the micelles at the molecular origin of life, formed from the lipids present on those snowball-like asteroids that produced the earth's primordial ocean. Such micelles were warmed by the sun during the day, causing them to liquify and deform; at night they would solidify and reestablish their original form due to hysteresis or "molecular memory." This is the origin of memory, which is the preadaptation that the phenomenon of brain cooling is in reference to.

Niche Construction as the Foundation for Social Morality

The recognition that the unicell was the first niche construction (Torday 2016) provides a mechanistic explanation for the integral physiologic-ecologic connections between individual members of an ecosystem (see Fig. 11.2). Given that moral behavior has evolved from physiologic principles based on quantum mechanics, the cell as niche construction would also explain why moral behavior transcends the individual, extending to other aspects of the environment, both animate and inanimate alike.

The social philosopher Murray Bookchin developed the concept of social ecology (Bookchin 1995), appealing to our sense of oneness with the world, giving rise to the environmental green movement that persists to this day, which also appeals to our innate sense of responsibility for ourselves, our environment, and the planet. What is being offered here is a scientifically based rationale for the same thing Bookchin was expressing intuitively, for if the cell and the atom are homologous, there is a true continuum from the singularity to life.

As empiric evidence for such a continuum in human terms, Nicholas Christakis, an epidemiologist at Yale, has shown empirically that there is a "network" of people within and between communities, which he refers to this phenomenon as "social contagion theory" (Christakis and Fowler 2013). But for the sake of this article, it demonstrates that humans practice niche construction in a manner consistent with the continuum from cell to self to community to global community (= Gaia).

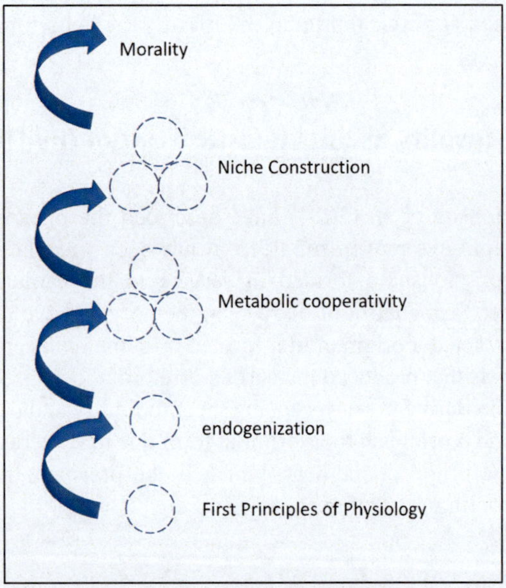

Fig. 11.2 The genesis of morality. Beginning with the first principles of physiology (bottom), endogenization, metabolic cooperativity, and niche construction, morality (top) is the anthropomorphic expression of those processes

Metabolic Cooperativity as the Origin of Biologic Morality

Life was represented by unicellular organisms for the first 3.5 billion years, followed by the rise of multicellular organisms some 500 million years ago. That transition may well have occurred due to competition between prokaryotes and eukaryotes, prokaryotes "inventing" such pseudo-multicellular traits as biofilm and quorum sensing. In response to that challenge, eukaryotes devised cell-cell communication using soluble growth factors and their cell-surface receptors to transfer energy. Such ligand-receptor interactions produce second messengers composed of high energy phosphates like cyclic adenosine monophosphate and inositol phosphate, which generate cascades that culminate in changes in DNA readout for growth and differentiation of the target cell. The cell-cell signaling of the embryo produces the offspring. Epigenetic "marks" obtained from the environment modify specific DNA codes within the genome, providing advanced notice of changes in the environment. The acquisition of such "marks" is accomplished by the phenotypic behavior of the offspring in its environment as a result of iterative cycles of epigenetic inheritance, which may eventually become constitutive. Such behavioral changes in response to the environment are tantamount to morality.

Phenotypic Agency and Moral Behavior

It is now well-recognized that Lamarckian inheritance directly from the environment is scientifically demonstrable. Epigenetic "marks" are acquired from the environment and transferred to the germ cells – egg and sperm – during the process of meiosis (Maamar et al. 2021). Subsequently, such "marks" are incorporated into the germ and somatic cells during embryogenesis, affecting the phenotype of the offspring.

Although the exact mechanism for selection of the marks remains undetermined, it is safe to say that it informs the organism of specific changes in the environment that need to be adapted to, affecting the phenotypic expression of the offspring, both somatic and behavioral, hence the connection between epigenetics and morality as a determinant of behavior. By definition, those behaviors that are adaptational would benefit the organism, including morality.

Morality Is in Our DNA

As stated earlier, life began as an ambiguity of energetic states, with negative entropy within the cell and positive entropy in the surrounding environment. Negentropy violates the second law of thermodynamics, heat dissipating over time, which does not apply in the case of living organisms, until they die. This state of cheating nature has been coped with by deception, of ourselves, of other organisms, and of our surroundings. The latter is a consequence, for example, of simple and complex machines, reducing the amount of energy expended to do work. The air foil, or wing, for example, allows for flight because of the differential in the rates of air passing over and under it, generating lift. For that matter, all physiologic properties comply with laws of physics. However, we are the only organism on earth that transgresses the boundaries of physics, wreaking havoc and hegemony.

Machines "cheat" nature, and all organisms do so to one degree or another, but humans have exploited machines to the point where the applications are unsustainable, causing them to ultimately collapse. In terms of universal homeostatic principles as the underpinning of moral behavior, such practices are definitionally immoral.

Morality in the Anthropocene

We are now in the Anthropocene, or an environment that is man-made, moving forward. Given that it is we who are forming the interrelationships, it behooves us to define morality so we know how to behave in a place of our own design and making. Recognizing the significance of homeostatic control as the basis for morality before the Anthropocene would seem to be a logical basis for formulating the moral code moving forward. For example, when the virtual reality program "Go to Meeting" was instituted, it seemed like a way to more efficiently facilitate interrelationships in virtual space. However, it was found that some participants were having their iconic representatives act in ways that they would not exhibit in real time. So rules of behavior had to be formulated. Similarly, if we are to function in a human-designed environment, we must determine the rules of behavior based on some set of principles. Given that conventional morality had its origins in homeostasis as the "equal and opposite force" that "balanced" the blast of the Big Bang, it would make sense to use homeostatic equipoise as the guiding principle for man-made morality as well. Stephen J. Gould famously asked whether we could "rewind the tape of evolution?" It is maintained that there are no "do-overs" for evolution. In that vein, deviating from the principles of evolution would be an existential mistake.

Altruistic Behavior in Bacteria

In his book *The First Minds*, Arthur Reber (2018) speaks of altruism in bacteria, the usual criterion for an act of altruism being that it benefits another while having a detrimental effect on the actor. Gurol Suel's study of bacteria shows that bacteria in one part of a colony adjust their metabolism and cell division rates in response to molecular signals from bacteria in other parts of the colony. The cells at the periphery of the colony enjoy access to the nutrient-rich environment but place themselves in jeopardy to predators, antibiotics, and toxins. The cells at the center enjoy the protection of their location but have diminished access to food. If the nutrient levels in the core of the colony drop too low, the cells there secrete a signaling molecule that essentially asks the cells on the edges to slow down, stop reproducing, and let some of the nutrient-rich material flow inward. When the cells in the periphery accommodate, they are, in effect, engaging in an altruistic act. They behave in ways that increase the welfare of their colony mates but have placed themselves in jeopardy. The potential dangers from predators and noxious molecules still lurk in the surrounding environment, and by lowering metabolic rate and decreasing cell division, they are diminishing their chances of passing their genes on to the next generation. The same phenomenon occurs between colonies too. And this behavior holds true both within and between different strains of bacteria.

Such behavior resembles kin selection, the evolutionary strategy that favors the reproductive success of an organism's relatives, possibly at the risk of the organism's own survival and reproductive success. This behavior has been well described and investigated, but its cause has been ascribed to increasing gene frequency. However, describing a mechanism is not the same as knowing how and why it occurs. On the other hand, the mechanism of niche construction (see above) would account for such behavior, using homeostatic balance as a measure of kin selection.

Discussion

Moral behavior is conventionally thought of in terms of religious precepts, made secular by laws. But in this article, it is argued that moral behavior is actually founded on biologic principals derived from natural laws. Using that logic, we would all behave morally, but because of our ambiguous origins between negative and positive entropy, within and outside of the cell, respectively, there has been a tendency to deceive both ourselves and others, necessitating the codification of laws, both statutory and religious. However, once it is realized that "being good" is the natural state of being, perhaps it is easier to understand the order of things.

In contrast to that, Darwinian evolution based on survival of the fittest mandates that organisms compete for reproductive success. Tennyson's phrase "Nature, red in tooth and claw" expressed this perception of how and why evolution occurs, but it is a gross distortion of the situation at hand. Yes, it is what catches our eye and attention, i.e., "if it bleeds, it leads," but like the debate between Darwinian "gradualism" and Eldridge and Gould's "punctuated equilibrium" (Eldredge and Gould 1972), the latter is the tip of the evolutionary iceberg of activity at a level that is invisible to the eye. Admittedly, mortal combat is far more interesting than the glacial changes that occur below the surface. And the metric for survival of the fittest is much easier to calculate than trying to understand the machinations of cells and molecules. But the reality is that it is the latter that is responsible for survival, by and large. And at that level, metabolic cooperativity is the order of the day, so being good comes naturally.

Axelrod and Hamilton had elegantly explained the evolution of cooperativity based on the "Prisoner's Dilemma," providing a rationale for such otherwise counterintuitive behavior in a survival of the fittest' context. However, such description is superseded by the mechanistic processes described herein, exemplifying the power of a cellular-molecular first principles approach to the question of evolution over a top-down Darwinian rationalization based on reproductive success as the metric for evolution.

The question of the origin of moral behavior is not just sophistry. In an age where we are considering such technologies artificial intelligence, it behooves us to understand how and why we behave. A debate is now raging as to what metrics to use in a situation where an autonomous vehicle must "decide" whether to kill a senior citizen or a mother with a baby carriage in its path, for example. Such decisions come down to material "cost," but what about the cost to society? Sure, the senior has outlived his earning power and the infant hypothetically has infinite earning power, but what of the contribution to family, community, and society? If our morals derive from our biology, there should be an algorithm that also accounts for such relationships, given the recognition of the organic nature of "being good." Such an algorithm was alluded to above in the context of kin selection as a corollary to value.

Such insight to moral motivation is an adjunct to the big history effort to understand our place in the cosmos. The combination of the cell as niche construction with Christakis's contagion theory acts as a unifying, holistic continuum for understanding human behavior, good, bad, and otherwise. As such, it would aid in understanding our deepest motivations, past, present, and future.

Morality and Donut Economics

In that vein, it is of interest that Kate Raworth (2017) has formulated her "donut economy theory" in which life is wedged between the biologic and the cosmologic. The model allows for the dynamic interrelationship between life and nonlife, economically based on cybernetic controls, not unlike the way we have envisaged physiology based on cell-cell communications (Torday and Rehan 2012). The net result of such a perspective would be the integration of economics with the cosmos, founded on the flow of free energy instead of supply and demand. Such a stellar interrelationship would also account for moral behavior being an innate quality of living organisms, answering Derek Parfit's question as to how to be good, referenced in the introduction.

References

M. Bookchin, *Philosophy of Social Ecology* (Black Rose Books, Montreal, 1995)

N.A. Christakis, J.H. Fowler, Social contagion theory: Examining dynamic social networks and human behavior. Stat. Med. **32**, 556–577 (2013)

D. Deamer, The role of lipid membranes in life's origin. Life (Basel) **7**, 5 (2017)

N. Eldredge, S.J. Gould, *Models in Paleobiology* (Freeman, Cooper, San Francisco, 1972)

J.A. Hobson, C.C. Hong, K.J. Friston, Virtual reality and consciousness inference in dreaming. Front. Psychol. **5**, 1133 (2014)

M.B. Maamar, E.E. Nilsson, M.K. Skinner, Epigenetic transgenerational inheritance, gametogenesis and germline development. Biol. Reprod. **30**, ioab085 (2021)

M. MacFarquhar, *How to Be Good* (The New Yorker, New York, 2011)

P. Mitchell, Coupling of phosphorylation to electron and hydrogen transfer by a chemi-osmotic type of mechanism. Nature **191**, 144–148 (1961)

S. Pinker, *The Better Angels of Our Nature* (Penguin, New York, 2012)

K. Raworth, *Donut Economics: Seven Ways to Think Like a 21st Century Economist* (White River Junction, Vermont, 2017)

A. Reber, *The First Minds* (Oxford University Press, Oxford, 2018)

J.S. Torday, V.K. Rehan, Lung evolution as a cipher for physiology. Physiol. Genomics **38**, 1–6 (2009)

J.S. Torday, The cell as the first niche construction. Biology (Basel) **5**, 19 (2016)

J.S. Torday, W.B. Miller Jr., The resolution of ambiguity as the basis for life: A cellular bridge between Western reductionism and Eastern holism. Prog. Biophys. Mol. Biol. **131**, 288–297 (2017)

J.S. Torday, V.K. Rehan, *Evolutionary Biology, Cell-Cell Communication and Complex Disease* (Wiley, Hoboken, 2012)

J.S. Torday, V.K. Rehan, *Evolution, the Logic of Biology* (Wiley, Hoboken, 2017)

Chapter 12
"Snookered"

"Snookered"

It's right there, in the lobby of the apartment building we've retired to,

It's Siren call beckoning me every time I stand there waiting for the elevator, pulled in opposite directions, just like it did back in college.

There I navigated the Scylla and Charybdis of the poolroom versus the classroom.

But I no longer need the pool table, "snookering me" into the meaning of life, for I have finally figured out how to use what I have learned as a laboratory scientist in the other classrooms to cope, verily to know, yes, I now understand the cosmos beyond the meta "Cosmos of the Snooker Table."

I spent many an hour playing pool during college, finding more than in my classes, sizing up the table lie, wondering if "I know that shot" (determinism), or have to take risk, the shot based on some emergent "combination" of data conjured up from my long-term memory under similar circumstances (probability).

Only years later did I realize that those two options were the soul of the atom and the unicell, life deriving from nonlife *symbiogenically*, as L. Margulis taught us.

And perhaps the cue ball is the unicell, imparting energy to the other balls to form and reform across the cosmos of the pool table, all under the rule of gravity.

Now I realize that it was because I thought I had more control over the knowledge.

I had gained on that baize felt, illumined from above as "episteme," the balls constrained by those finite, firm, resilient green cushions, meaning more than English, biology, physics, chemistry, sociology, and psychology combined.

It would take many many many years to sort that out, finally able to realize that our physiology is the matrix for the laws of nature, common to both our consciousness and the consciousness of the cosmos. Krishnamurti teaches that we can reach that cosmic consciousness by foregoing our ego, flipping a switch, but the scientist in me cannot do that, choosing instead to reach that goal stepwise as a rheostatic switch for ontogeny and phylogeny.

J. S. Torday, *Hormones and Reality*, https://doi.org/10.1007/978-3-030-93691-4_12

Now, in my eighth decade, I realize both the determinism and probability of that next pool shot is Newtonian, whereas what I had conjured up from my study of cell biology was Planckian quantum mechanical.

You cannot experience QM on a pool table, but it gives onto the portal for asking "how and why," my mind caroming around the pool table, entering the periodic table of elements at one end, exiting it as a biologic algorithm at the other, finally recognizing the significance of both the synchronic "now" and the diachronic "across space-time" aspects of the elements that Mendeleyev gifted us.

I doubt he knew that he had set in motion the possibility of regressing the alchemical physicality of the elements against their proton numbers, producing Qbits for quantum computing,

The first cell providing a crucible for the resolution of quantum entanglements, referencing the nonlocal events of the cosmos, just as they had formed the planets, offering an explanation for Jung's synchronicities.

Eureka! At long last, the final common denominator for all of knowledge as the consciousness of the cosmos.

Deep within us, all along, "the secret sits."

Chapter 13
Mr. Bubble Creates Civilization

The Bubble family was frolicking in the surf one summer's day, when out of the blue Bubble junior asked his father "Where did we come from?" To which his father replied that "Once upon a time, deep in space, your great great great grandparents came from Pulsaria and landed on the water that covered the planet earth. Because we bubbles are made of electrically charged heads and tails, we lined up in rows vertically to the water, like soldiers in formation, and reduced the surface tension of the water, like when water in a straw creeps up the sides, so they could remain together as a family. Then the force of gravity that keeps the bubbles from floating off into space caused those rows of lipids to form bubbles. But they weren't ordinary bubbles. Their surfaces had tiny invisible holes in them that allowed particles to enter and exit the bubbles freely, under the influence of physical principles like quantum entanglement and non-localization that cannot be seen or felt, but they are there. Such principles connect the bubbles to the cosmos.

Because the bubbles floated at the surface, they were warmed by the sun during the day. As a result, the bubbles liquified, and so they got bigger, letting in more and more minerals dissolved in the water like calcium and sodium. Too much calcium could cause the bubbles to die. But some bubbles figured out a way of controlling the amount of calcium that passed in and out of them. Those are the bubbles that are our ancestors.

There were bacteria gangs in the neighborhood too, who bullied the bubbles. One day the bacteria figured out ways of cooperating with one another, allowing them to gang up on the bubbles and bully them. But the bubbles were smarter, and they developed their own superior ways of working together through communications, allowing them to fight off the bacteria.

Some bubbles became algae, floating on the surface of the water, producing carbon dioxide, the gas in your coca cola. Eventually carbon dioxide built up in the air, creating a "greenhouse effect" that warmed the air in the atmosphere up, causing some of the water to dry up and expose the underlying land.

J. S. Torday, *Hormones and Reality*, https://doi.org/10.1007/978-3-030-93691-4_13

The increased heat in the atmosphere also caused a decrease in oxygen in the water, driving fish out of the water onto land. The fish that were able to adapt to land had a so-called swim bladder, like "swimmies," that helped them float in the water. On land, the swim bladder evolved into a lung like yours and mine, allowing oxygen in the air to enter the fish to digest food.

Every now and then, the evolving lungs were not big enough to allow food to digest, causing stress on the fish-men, stimulating the production of the fight-or-flight hormone adrenalin. The adrenalin caused the tiny air sacs in the lung called alveoli to expand, allowing in more oxygen for energy. It also caused the release of food from fat cells, which could be digested very well for energy, increasing body heat.

Eventually, genes evolved for the constant control of body heat, allowing the fish-men to stand up on their hind legs, freeing their front legs for special activities such as making tools and speaking and writing words or language. At first it was speaking out loud, and later there was written language by combining toolmaking with language.

Written language ensured the faithful communication of culture within and between generations. Oral language encouraged people to live in small groups, exchanging knowledge about hunting and gathering.

Over time, the groups got larger and larger, forming tribes, colonies, states, and nations. Knowledge was passed from generation to generation using clay tablets for handwriting, and eventually the printing press was developed, allowing for the production of printed information that could be shared with large numbers of people both within and between generations. Eventually, such large groups developed cultural activities such as schools for learning, art institutes, music, schools, governments, etc. culminating in civilizations.

Chapter 14
Like History, Evolution "Rhymes"

Check for updates

Introduction

Since evolution represents the biologic "history" of man, it should be functionally integrated with our social history. Furthermore, conventional Darwinian evolution is not mechanistic, allowing for a testable/refutable understanding of the causal relationships underlying the process. By merging history with "non-machine like" mechanistic evolutionary biology, the ultimate goal of history would be realized.

The environment has literally molded life on earth from the very start. The almost metaphysical formation of micelles, or protocells, in the primordial oceans set the process of evolving life in motion. Subsequent production of carbon dioxide by plants caused its accumulation in the atmosphere, generating a "greenhouse effect," warming the atmosphere, partially drying up the ocean covering the surface of the earth. That warming effect also caused a decrease in the oxygen content of the water, forcing physostomous boney fish, which have a trachea-like pneumatic duct connecting their esophagus to their swim bladder, onto land. In adapting to land, specific self-engineered gene duplications occurred during that transition from water to land.

Later still, during the Phanerozoic period, comprising the Paleozoic, Mesozoic, and Cenozoic, oxygen levels in the air varied between 15% and 35% (Berner 1999). The increases in oxygen caused gigantism in many species, whereas the subsequent decreases caused physiologic stress due to hypoxia. The hypoxic stresses were hypothesized to have given rise to endothermy-homeothermy by stimulating adrenalin production through the hypothalamic-pituitary-adrenal axis. Endothermy-homeothermy, in turn, gave rise to bipedalism, freeing the forelimbs for specialized functions (flight in birds, toolmaking, and language) and higher consciousness (Torday 2015). The latter is critical to the concept of written history.

Therefore, the articulated integration of evolution with environmental change inextricably weds biology to history causally. That realization offers the opportunity

J. S. Torday, *Hormones and Reality*, https://doi.org/10.1007/978-3-030-93691-4_14

to further probe the depths of such interrelationships than would otherwise be possible by merely citing these processes as associations and correlations, thus gaining far deeper insights to history in the same vein as Jared Diamond's *Guns, Germs, and Steel* (1997).

The "Roll" of Evolution in History

Life on earth has been formulated by adaptive interactions between the organic and inorganic, mediated by evolutionary biology. The first links between the environment and life were those of animism and astrology, the latter segueing into Heliocentrism fostered by astronomy as the catalyst for the Age of Enlightenment. Subsequently, the redshift (Penzias and Wilson 1965) and the Big Bang theory (Woolfson 2013) have given us a narrative instead of circular reasoning, offering a logic to our understanding of our physical origins as a point source (Torday and Miller 2016).

Darwin himself had an intuition about the relationships between the environment and evolution, as expressed in his *On the Origin of Species* (Darwin 1859). He commented on the topography of Patagonia in great detail but never developed the idea further for his theory of evolution. Conversely, Lamarck recognized the intimate relationship between the environment and evolution but did not have the science needed to demonstrate the principle. Only recently has epigenetic inheritance come back into vogue, offering the opportunity to recognize the interrelationships between history and evolutionary biology.

History traces its arc from the Big Bang to the present as a continuum. The rationale for Big History is spelled out in Rodrigue et al.'s *Our Place in the Universe* (Rodrigue et al. 2016). The book is an introduction to the idea that the "story of everything" can be told, but in order to make it comprehensive, its physical and biologic aspects must be merged as functional elements of the totality. At its largest scale, Lovelock's Gaia theory (Lovelock 1972) and Smolin's application of Darwinian evolution to astronomy (Smolin 1999) have established the organic nature of earth and the cosmos, respectively. At its smallest scale, atomic theory and biologic cell-cell communication provide mechanistic consilience between quantum mechanics, the first principles of physiology, and evolutionary biology at the unicellular level. The merging of physics and biology within the cell (Margulis 2011) has offered the opportunity to consider the congruence of the inanimate and the animate, referring all the way back to the singularity/Big Bang based on empiric evidence for the first time. The vertical integration of those principles has been exploited to explain the mechanism of physiologic evolution, allowing a rationale for incorporating the latter into the concept of Big History.

History and Consciousness

Biology would have no history if we were not conscious of our own existence. But what is consciousness? Is it "all in our heads" or "theater of the mind." Hameroff and Penrose's "Orch-Or theory" (Hameroff and Penrose 2014) has proposed an elegant physiologic explanation for consciousness as the networking of neurons through microtubules. But then again, all cells have microtubules in their cytoskeletons, opening up to the concept of consciousness as awareness of our whole body, referred to as allostasis. Alternatively, it has been hypothesized that consciousness is the essence of our physiology, based on stepwise principles of cell-cell communication. Experimental evidence for this derives from observing patients recovering from general anesthesia undergoing the phylogenetic steps of brain evolution from reptilian to mammalian (Mashour and Alkire 2013). Conversely, when lung or bone cells are experimentally exposed to microgravity, they lose their capacity to signal with the environment or with one another (Torday 2003). These data point to the fundamental nature of consciousness as the way in which organisms interrelate with the cosmos, given that gravity was a product of the Big Bang.

Thinking about consciousness in the above terms is advantageous because it is derivative of the Big Bang, integrating the inanimate with animate as a functional whole. Instead of anthropic principle, man being "in" the cosmos, we are "of" the cosmos, literally.

The ultimate purpose for considering evolution in the context of history is to raise our consciousness. Such philosophers as the pre-Socratics or scientists such as E.O. Wilson (1999) have attempted to do just that, but they lacked a core mechanism like the one being forwarded herein. Cosmic awareness is implicit in the codification of history; it is explicit in cellular-molecular evolution, emanating from the Big Bang stepwise based on cell-cell communication as a continuum from the origin of life itself.

In the Beginning

The earth formed 4 or 5 billion years ago. Initially, it had no atmosphere, allowing snowball-like asteroids to strike the surface and melt, forming the ocean that covered the earth. There were lipids embedded in the ice forming those asteroids, which spontaneously formed micelles or prototypical cells. The hypothetical lipid origin of life on earth makes both a priori sense because lipids exhibit hysteresis or "molecular memory" necessary for the process of evolution and a posteriori because Jack Szostak has shown experimentally that lipids can synthesize nucleotides, but nucleotides cannot synthesize lipids. The semipermeable membrane micelles offered a protected space for the first principles of physiology – negentropy, chemiosmosis, and homeostasis (Torday and Rehan 2009).

Endosymbiosis Theory

Ivan Wallin was the first to propose endosymbiosis theory, which was popularized by Lynn Margulis Sagan. She asserted that eukaryotic cells, with nuclei, are products of the symbiotic partnership with bacteria. It is now well accepted that the cellular mitochondria that are crucial for cellular energy metabolism were previously free-living bacteria that became inherent components of the eukaryotic cellular apparatus. The core concept is that eukaryotes evolved by endogenizing environmental factors that posed existential threats over the course of vertebrate history. In this context, history complements our understanding of physiologic evolution by providing the sequence of changes in the environment, both natural and man-made, that have affected our evolution. And since evolution is the history of biology as serial preadaptations or exaptations (Gould and Vrba 1982), it helps to gain a deeper understanding of the course of human evolution.

Cell-Cell Communication as Physiologic Evolution

Epigenetic inheritance consists of obtaining data from the environment in combination with cell-cell communication, developmentally, phylogenetically, and homeostatically. That integrative relationship between the organism and its environment is what we refer to as evolution; a disconnect between them causes physiologic stress in the tissues and organs being affected, producing reactive oxygen species (ROS) (Storr et al. 2013). Such ROS causes tissue-specific gene mutations and duplications, which can be resolved by adaptation as evolution. Short of reengineering physiologic traits, injuries heal based on the same homeostatic principles of cell-cell communication.

Endosymbiosis: The Laws of Nature and Consciousness

The cell authors its own internal "laws" based on the homeostatic laws of nature, formulated by Claude Bernard as the *milieu interieur* and by Walter B. Cannon as homeostasis. Taken together, cellular homeostases are referred to as allostasis in the aggregate, monitored and controlled by the nervous system. This holistic physiologic process of self-awareness is what we conventionally think of as consciousness.

The Cell as the First Niche Construction Is a Continuum Between Man and Environment

Niche construction refers to the process by which organisms actively fashion their immediate environment to optimize their adaptation. This phenomenon was first observed by Darwin himself, noting that earth worms retain their aquatic kidneys on land. As odd as that may seem, it is what endosymbiosis theory constitutes, internalizing factors in the environment that have posed threats to their existence, beginning with the unicellular state, which can be seen as internal niche construction. The concept of the *milieu interieur* was later codified by Walter B. Cannon as physiology. Ultimately, the internalization of physical factors functioning under the laws of nature conferred this property on life, forming organic ties between the cell and the environment as a continuum from the unicell to Gaia. As a result, history could be thought of as the description of this process, whereas understanding the underlying mechanisms that causally link the organism to its environment adds innate depth to the process. And it offers the opportunity to recognize transcendent interrelationships, reaching into the interstices of the cosmos to gain fundamental understanding of the process. Such an exhaustive analysis lends itself to finding common ground between eastern and western philosophy, tearing down the barriers of contemporary knowledge to maximize the accounting of history.

Top-Down, Bottom-Up, Middle-Out

Biologic control is referred to as either top-down, bottom-up, or middle-out. Top-down control references emergence of physiologic properties. Bottom-up references the organization of physiologic traits from its component parts. Middle-out is the result of cell-cell communication through growth factor-growth factor receptor signaling.

Combining Epigenetic Inheritance with Phenotypic Agency Provides Biologic Scope to History

Epigenetic inheritance constitutes the collection of epigenetic marks over the course of the life cycle. The so-called marks are then integrated into the DNA of the germ cells (egg and sperm) as adducts (methylation, ubiquitination, myristoylation, etc.) that modify the nucleotide "readout" in accord with environmental changes. The DNA adducts subsequently appear in specific tissues and organs, where they modify the structure and function of the organisms accordingly as epigenetic inheritance.

Rather than the phenotypes of the adults determining evolution, as dictated by Darwinian evolution, the central role of germ cells in epigenetic inheritance infers the primacy of these cells in the process of adaptation. In that light, the phenotype can be viewed as an active "agent" for the acquisition of epigenetic marks, the individual adopting an active role in history based on the biologic imperative of acting as a vehicle for epigenetic inheritance.

Being "In" This Cosmos (Anthropic Principle) Versus Being "Of" the Cosmos (Endosymbiosis)

The anthropic principle is the concept that we fortuitously ended up in this particular place in the cosmos. Conversely, evolution has facilitated our adaptation to our environment by endogenizing it, rendering what otherwise would have destroyed us billions of years ago – gravity, oxygen, heavy metals, and ions – useful as what we now recognize as our physiology. For example, one can see the causal relationships involved by regressing the genes that facilitated lung evolution against major epochs in the geochemistry of the earth (Torday and Rehan 2011). On the other hand, as Jean Guex has demonstrated using ammonites, environmental stress can disrupt and reverse the evolutionary process (Guex et al. 2020).

Evolution as the Underlying Mechanism of History

Merging history with evolutionary biology tacitly acknowledges that both originated in the Big Bang. This perspective confers a deep understanding of who and what we are historically, and evolutionarily offers the organismic context for history.

Conclusions

History rhymes because the underlying process of evolution "rhymes," founded on serial preadaptations or exaptations, i.e., organisms reappropriate genetic motifs that were effective at some earlier stage in their evolutionary history when confronted with an existential threat, ultimately referencing the first principles of physiology – negentropy, chemiosmosis, and homeostasis. Such principles, in turn, reference the Big Bang as their origin. So like Big History, biology is also the product of the Bing Bang. By recognizing the homologies between the two processes, we can better understand the human condition from its source rather than reasoning after the fact.

There are certain narratives such as those of the Old and New Testaments, the golden rule, the US Constitution, the scientific method, the periodic table, and the laws of nature that have served us well. We have inherited certain laws of nature biologically that have served us well throughout our evolutionary history. If we were to understand the absolute interrelationships between such principles, we would optimize history.

The ancient Greek philosopher Protagoras said that "Man is the Measure of all things"; he was right in spirit, but he needed to know what the "units" of measurement were to support his idea scientifically. For the scientist, it is the cell. For the humanist, the cell is the "syntax" of Big History. The cell as the mechanistic basis for both evolution and Big History offers a novel synthesis for humanism and science, bringing resolution to C.P. Snow's "Two Cultures."

In his Big History, David Christian references the "goldilocks" effect explanation for our fortuitous existence, describing the mechanism of homeostasis, without which neither the inanimate nor the animate can exist. Harold Morowitz describes how the electron and proton balance one another energetically within a hydrogen atom. And in the cellular-molecular approach to evolutionary biology, homeostasis is one of the three principles of physiology, controlling the interrelationship between negative entropy and chemiosmosis. The Pauli exclusion principle and the first principles of physiology are both deterministic and probabilistic, offering the opportunity for stability and plasticity.

References

R.A. Berner, Atmospheric oxygen over Phanerozoic time. Proc. Natl. Acad. Sci. U. S. A. **96**, 10955–10957 (1999)

C. Darwin, *The Origin of Species* (John Murray, London, 1859)

J. Diamond, *Guns, Germs and Steel* (WW Norton, New York, 1997)

S.J. Gould, E. Vrba, Exaptation: The missing term in the science of form. Paleobiology **8**, 4–15 (1982)

J. Guex, J.S. Torday, W.H. Miller Jr., *Morphogenesis, Environmental Stress and Reverse Evolution* (Springer, New York, 2020)

S. Hameroff, R. Penrose, Consciousness in the universe: A review of the 'Orch OR' theory. Phys Life Rev **11**, 39–78 (2014)

J.E. Lovelock, Gaia as seen through the atmosphere. Atmos. Environ. **6**, 579–580 (1972)

L. Margulis, Symbiogenesis. A new principle of evolution rediscovery of Boris Mikhaylovich Kozo-Polyansky (1890–1957). Paleontol. J. **44**, 1525–1539 (2011)

G.A. Mashour, M.T. Alkire, Evolution of consciousness: Phylogeny, ontogeny, and emergence from general anesthesia. Proc. Natl. Acad. Sci. U. S. A. **110**, 10357–10364 (2013)

A.A. Penzias, R.W. Wilson, A measurement of excess antenna temperature at 4080 Mc/s. Astrophys. J. Lett. **142**, 419–421 (1965)

B. Rodrigue, L. Grinin, A. Korotayev, *Our Place in the Universe* (Primus Books, Delhi, 2016)

L. Smolin, *The Life of the Cosmos* (Oxford University Press, Oxford, 1999)

S.J. Storr, C.M. Woolston, Y. Zhang, S.G. Martin, Redox environment, free radical, and oxidative DNA damage. Antioxid. Redox Signal. **18**, 2399–2408 (2013)

J.S. Torday, Parathyroid hormone-related protein is a gravisensor in lung and bone cell biology. Adv. Space Res. **32**, 1569–1576 (2003)

J.S. Torday, A central theory of biology. Med. Hypotheses **85**, 49–57 (2015)

J.S. Torday, W.B. Miller, The unicellular state as a point source in a quantum biological system. Biology (Basel) **5**, 25 (2016)

J.S. Torday, V.K. Rehan, Lung evolution as a cipher for physiology. Physiol. Genomics **38**, 1–6 (2009)

J.S. Torday, V.K. Rehan, A cell-molecular approach predicts vertebrate evolution. Mol. Biol. Evol. **28**, 2973–2981 (2011)

E.O. Wilson, *Consilience* (Vintage, New York, 1999)

M. Woolfson, *Time, Space, Stars & Man: The Story of Big Bang* (Imperial College Press, London, 2013)

Chapter 15
On the Evolution of Imagination as Human Consciousness or "Imagining Imagining": A "Parsimonious" Perspective on Imagination and the Evolution of Human Consciousness

"Once there were bacteria, now there is New York" was penned by Simon Conway-Morris, emblematic of our ability to conceive of our own origins and arc of evolution. By imagining imagining, we can finally "see" its origin as a "by-product" of endothermy, physiologically determined by the evolution of the hormone oxytocin, which took over from the conditional stress reaction to intermittent hypoxia caused by the evolution of the lung from the swim bladder of boney fish during the transition from water to land. Seen in the forward direction, this feature of humans is the product of warm-bloodedness affording a relatively constant body temperature, rendering our metabolism efficient for standing on two legs, or bipedalism, freeing our forelimbs for toolmaking, including language as a "tool" (Brozzoli et al. 2019). That convergence of freedom of range due to endothermy in combination with toolmaking and language, particularly written language, forged human civilization as the repository and catalyst for imagination. And just to clarify, other organisms also express oxytocin, but it is in combination with freeing of the forelimbs that catalyzed imagination, which makes being human possible, including composing this chapter.

I have been thinking about the evolution of human consciousness a great deal. It has occurred to me that perhaps it is a consequence of terminal addition (Torday and Miller Jr 2018), the process of adding novelty onto a series of evolved traits, like fingers and toes at the ends of hands and feet, for example, or cellular changes that progressively facilitate evolution of internal organs, like the lung alveoli and kidney glomeruli. With that idea in mind, I kept wondering what the terminal addition to garden variety consciousness – what it is like to be a bat (Nagel 1974) – would constitute human consciousness, and then it struck me that perhaps its imagination. And the more I thought about it, the more it seemed to resonate with the way I have been thinking about the evolution of physiology, as emergent properties coping with existential problems (Torday and Rehan 2012). So in other words, imagination is a rapid, nonlife threatening way of producing emergence as a way of problem-solving.

J. S. Torday, *Hormones and Reality*, https://doi.org/10.1007/978-3-030-93691-4_15

Just as a further thought that hypothetically interconnects physiology, evolution, imagination, and creativity, when David Chalmers talks about "the hard question" as to why we see "red" when we injure ourselves (Chalmers 1995), I think that is a manifestation of the way I am thinking about imagination from the inside-out as our capacity to integrate physiologic properties as the "muse" for creativity. And because our physiologic "insides" are the environmental "outsides" based on endosymbiogenesis theory/developmental physiology, it is the synergy between environmental cues and our physiology that is the inspiration for imagination and creativity. So for example, music literally "resonates" with our physiologic being, forming and/or reinforcing pathways viscerally and neurologically, facilitating imagination.

In a previous article entitled "A Central Theory of Biology" (Torday 2015b), the evolution of the human central nervous system was traced from the advent of endothermy to bipedalism, the latter freeing the forelimbs for toolmaking and language as a derivative of toolmaking. In our minds, the process of forming a sentence is no different from fashioning a spear tip out of flint, taking a piece of stone as our "subject," carving it as a verb, and producing the desired "object" as a spear tip. That sequence of evolutionary events placed positive selection pressure on the primitive hominin central nervous system (CNS), which is principally composed of lipids, referring all the way back to the alignment of the lipids in the air/water interface produced by pulsars and hydrothermal vents in the seafloor. When lipids coalesce at the water surface, they align perpendicularly, packing together, reducing the surface tension of the water, allowing for the formation of micelles as prototypical cells. This is in contrast to plant cells, which have a stiff cellulose cell wall that provides stability. Perhaps that is why they are immobile and orient downward as their gravitropism, having their conscience in their roots according to Frantisek Balushka et al. (2009).

Unicellular life dominated the earth for 3.5 billion years. Subsequently, the organizational principle underlying the process of multicellular evolution entails terminal addition, newly acquired traits literally being added on to the ends of a series of adaptive traits, like the fingers at the ends of our arms. The underlying mechanism for this is that physiology is composed of cells that communicate with one another by binding soluble growth factors with their receptors, triggering the formation of high energy phosphates that stimulate DNA readout for either growth or differentiation. Inserting a new trait in the middle or at the beginning of such a signaling sequence would subvert the ultimate purpose of evolution to perpetuate the first principles of physiology – negative entropy, chemiosmosis, and homeostasis, which reference the singularity of nature as the origin of the cosmos, inanimate and animate alike.

Regarding the positive selection for the central nervous system, further evolution of the hominin brain was due to freeing of the forelimbs for toolmaking and language, language being a homologue of toolmaking. Both toolmaking and language are localized in the area of Broca, the Foxp2 gene promoting synergy between the two (Xu et al. 2018). That synergy forged written language as the convergence of toolmaking and language, followed by the printing press, disseminating knowledge much more broadly and consistently than ever before from one generation to the next, ensuring faithful transmission of accumulated, buildable knowledge. The ultimate product of such techno-linguistic advancement was Western civilization, beginning in the Fertile Crescent, followed by Greece, the Roman, the Ottoman and Holy Roman Empires, Europe, and the United States.

The complementarity between the individual and civilization cited above is homologous with the process of niche construction, the unicell being the first biologic entity to endogenize its environment in a serial manner, forming patterns and networks within and between organisms that we refer to as ecologies; yet those ecologies are ultimately formed based on the same set of natural laws as is the cosmos (Torday 2020). This is much the same as David Bohm's "enfolding and unfolding." The enfolded order is the implicate order, where space and time are no longer the dominant factors that determine the relationships of dependence or independence of different elements. Instead, our conventional notions of space and time, along with those of separately extant material particles, are abstracted as forms derived from the deeper asymmetric interrelationships initiated by the Big Bang, which are resolved by physics, chemistry, and biology alike. Such ordinary notions in fact appear in what is called the explicate or unfolded order, a special and distinguished form contained within the general totality of all the implicate orders.

Bohm teaches us that the explicate order was formed by our subjective senses, which have allowed us to adapt to our environment, bearing in mind that there is an ideal implicate order (Bohm 1980). The way in which we transcend the explicate order and approach the implicate order is through experimentation, either formally using science or informally through our occupation with the arts – literature, sculpting, painting, and music – sport, and introspection. More specifically, all of these activities reflect our internal mapping of the cosmos through the first principles of physiology, referencing the singularity. That "map" has evolved through the cell-cell communications that form the basis for multicellular organisms, controlled by homeostasis as the ongoing manifestation of the "equal and opposite reaction to the Big Bang" (Torday 2015a). In the case of physics and chemistry, the homeostatic principle culminates in stable, static structures, whereas in the case of biology, founded on negative entropy, it is in a perpetual state of ambiguity, so it is the antithesis of stasis.

Imagination and creativity are embedded within us biologically as the first principles of physiology – negentropy, chemiosmosis, and homeostasis – the latter as the force that maintains stability. But when homeostasis is perturbed, it causes the system to reach back into its self-referential, self-organized history to identify an emergent trait that had proved useful earlier based on similar physicochemical and biologic principles and will now do so again, faced with yet another inevitable existential threat, given the ever-changing environment. It is that emergent quality which we think of in terms of imagination and creativity in wonderment. In fact, this phenomenon is consistent with the autopoietic nature of life first described by Maturana and Varela as a metaphor without a mechanistic explanation (Maturana and Varela 1980). If it were not for the exaptational, preadaptive character of this process, referencing the unicell as the origin of life, imagination and creativity would not manifest themselves.

Ultimately, the unicell references the singularity that existed before the Big Bang; so looking in the prograde direction, it is the singularity that is the author of the inventive, imaginative, and creative force behind the cosmos.

As was stated elsewhere, the atom and the cell are homologues (of the same origin), both having deterministic and probabilistic features. It is that fundamental kinship, facilitated by the endogenization of factors in the environment such as gases, matter, and prokaryotes (bacteria) that forged the evolution of life from non-life, perpetuated by the momentum of the Big Bang, homeostasis being the equal and opposite force that gave rise to the matter that was derived from the energy of that cataclysmic event. The immersion of lipids in water was the key to life as we know it on earth. Robert Frost stated in one of his notebooks that "Life is that which can mix oil and water" (Thompson 1966). As mentioned earlier, such lipids are produced by pulsars in deep space and by hydrothermal vents in the ocean floor. And to recapitulate, because they are amphiphiles, or zwitterions, with charged ends, they align at the air/water interface, pack together, and reduce the surface tension of water, making the environment conducive to the formation of primitive cells or micelles composed of spheres with semipermeable cell membranes.

Unicellular organisms dominated the earth's environment for 3.5 billion years. It has only been over the course of the last 500 million years that multicellular organisms have evolved in reaction to the existential threat by prokaryotes, engaging pseudo-multicellular forms such as biofilm and quorum sensing. In response, eukaryotes (our ancestors) began cooperating with one another metabolically in order to survive the prokaryotic onslaught.

Now, fast forward to imagination as the exaptational evolution of the central nervous system, it is the problem-solving feature that is available to us in lieu of "fight or flight," granted by the protection of civilization, with its resources, both material and psychological, fostering more and more of the same, given that civilization provides the "safe haven" in which to perpetuate such niche construction through self-perpetuation, servo-ed to the generative and regenerative properties of the cosmos.

The above way of understanding the evolution of consciousness as the "mapping" of our physiology onto the cosmological laws of nature resolves David Chalmers' "hard problem" – why we see "red" when we injure ourselves. It is a manifestation of our imagination, reprising the stepwise physiologic process of wound healing, bloody bits and all, elicited by the stress of pain as a trigger for reference to deeply atavistic traits. It also resolves Andy Clark's "extended memory" by merging the networked physiologic underpinnings of the human mind with civilization as niche construction, forged by the convergence and synergy between toolmaking and language.

Atavisms in biology have long been recognized anecdotally as the reappearance of ancient traits, like humans born with tails. And the linkage between such reappearances and developmental biology have similarly been documented. But with the recognition that pattern formation during development is due to cell-cell communication has opened up a new way of thinking about atavisms at the molecular level as recombinations and permutations, particularly with regard to internal organs like the lung and kidney. The lung is now known to have evolved from the swim bladder of boney fish, the first 50 genes involved in swim bladder development being identical with those in the lung (Zheng et al. 2011). So the utility of the swim bladder as an organ of gas exchange for adaptation to gravity as buoyancy in water has "morphed" into oxygenation for metabolism in terrestrial organisms. The underlying mechanism for such reconfigurations has long been recognized as neoteny or heterochrony, but in descriptive biology those processes have merely been seen anecdotally as the retention of juvenile phenotypes.

However, now with the benefit of being able to see the molecular underpinnings of the swim bladder and the lung, the how and why of the evolution of the former into that of the latter can be understood. Not only that, such comparisons lead to the deepest understandings of causation independent of physical appearances. For example, experimental evidence that the hormone leptin plays a role in *Xenopus* lung development led to the realization that the earliest principle involved in lung evolution from the foregut was not a function of air breathing but host defense, since leptin is a member of the cytokine interleukin-6 family of genes. The expansion of the foregut surface area as the primordial lung was facilitated by IL-6 acting in defense of this enlarging surface, given that the foregut is exposed to the environment. That relationship evolved into the implementation of leptin for alveolar evolution. In turn, that interrelationship references the skin as the most ancient organ of gas exchange, utilized by frogs for breathing. And the cellular-molecular homologies do not end there. The mechanism for host defense in the skin is homologous with that of the lung; the packaging of lipids that are excreted to form the stratum corneum is molecularly similar to that of the alveolus, surfactant proteins A and D being members of the defensin family of antimicrobial peptides.

This homology between the lung and skin with regard to host defense emerges with the etiology of asthma in dogs, for example (Machado and Ottolini 2015). Dogs develop asthma due to a polymorphism for β defensin CD103 that also determines coat color in the skin. This relationship demonstrates the hierarchical nature of such a chronic disease – coat color is critical for reproductive success in dogs, so it trumps the unfortunate effect of the same genetic trait in the upper airway causing wheezing. Such pleiotropic mechanisms are commonplace. In humans, the utility of type IV collagen to prevent epithelial "leakage" in the course of adaptation to land has resulted in Goodpasture's syndrome. A specific isoform of type IV collagen evolved under selection pressure for a stop-gap mechanism to prevent leakage across the lung alveolus and kidney glomerulus. That isoform resulted in an autoimmune reaction that causes Goodpasture's syndrome (MacDonald et al. 2006), the antibody attacking those sites, causing death in some cases. This disease occurs in a subset of people who have largely benefitted from the utility of type IV collagen as a means of preventing leakage of bodily fluids and formed elements in the blood. So evolution can be thought of as existentialism, coping with the absurdity of life due to our ambiguous origins as oil mixed with water.

Up until now, such neotenic relationships have been seen as anecdotal. The classic example is *Turritopsis dohrnii*, the so-called "immortal" jellyfish (Piraino et al. 1996). It is seen in this light because when it experiences stress it reverts back to its juvenile form, suggesting that it has discovered a way to immortalize itself when in reality it is merely taking on that form to obtain epigenetic marks using an atavistically tried and true phenotype.

The water-land transition, brought on by plants producing carbon dioxide, causing a "greenhouse effect," partially drying up land masses, and depleting oxygen from the water surroundings, forced boney fish onto land. In adapting to land, there were three known receptor gene duplications – parathyroid hormone-related protein, glucocorticoid, and β-adrenergic – all of which are critical for terrestrial life. Homing in on oxygenation, the lung was essential for land life, evolving from the swim bladders of boney fish in a stepwise fashion mediated by the remodeling of cellular structure and function in order to form alveoli from the air sac used for buoyancy as the swim bladder. The conversion of the organ from adaptation to gravity to oxygenation is an example of "imagination," fostered by the original formation of lipids at the air-water interface, providing orientation to electromagnetic forces like gravity. Importantly, over the 541 million year course of the Phanerozoic epoch since the Cambrian, oxygen levels in the atmosphere have fluctuated between 15% and 35%. The decrease in oxygen levels would have caused hypoxia, which is a potent stimulus for the production of oxytocin, hypothetically causing the evolution of human imagination.

In principle, it was the formation of the micelle that gave rise to the explicate order as the boundary between the internal space within the cell and the implicate order. That relationship is consistent with non-localization as a quantum mechanical principle.

The significance of imagination for problem-solving to formulate hypotheses without actually having to physically act is an alternative to the "fight-or-flight" response. It is that ability to problem-solve as a third option that distinguishes us from all other species.

Introspection is also a manifestation of imagination as the ability to self-analyze as self-referential self-organization. The capacity to "reflect", perhaps like mirror neurons (Asma mentions such in his book on the Evolution of Imagination (Asma 2017).

Think about the implications of "Out of Africa," going north, adapting to colder and colder climate through oxytocin, convergently acting to promote imagination. Is this hypothesis supportable? Are people in colder climes more imaginative or inventive than those in warmer environments? Had our ancestors in the Rift Valley invented the printing press, would the world be a "kinder, gentler place?" The premise of the book The WEIRDEST People in the World (2020) is that Western Europe began with kinship, followed by the Roman Catholic Church, the Protestant reformation, the printing press, and the Protestant work ethic. Anecdotally, the wine press was the forerunner of the printing press technologically. So the habit of drinking dilute wine instead of water saved Gutenberg's ancestors from dying during the plague years, only to come up with a way to disseminate knowledge faithfully from generation to generation, i.e., Western Civilization. So is that progression what gave rise to Western Civilization?

The comparative evolution of the forelimbs of humans versus birds is instructive in the context of the purpose of imagination. Only mammals and birds are warm-blooded. And humans and birds are the only bipedal organisms, freeing the forelimbs for special traits – flight in the case of birds and toolmaking/language in the case of humans, language being a form of toolmaking. Ultimately, the conditional generation of warm-bloodedness due to physiologic stress of hypoxia causes adrenaline secretion by the adrenals, causes lysis of fat cells, releases free fatty acids into the circulation, provides fuel for increased metabolism, and generates body heat. Ultimately, this conditional increase in body heat is replaced by oxytocin genetically determining body heat. In addition, oxytocin promotes imagination/creativity/ingenuity. Oxytocin is produced all the way back in vertebrate evolution to fish, but it is because oxytocin is produced in combination with freedom of forelimbs that it generated imagination/creativity/ingenuity for toolmaking and language in man.

But starting from the hypothesis that the phenotype is an agent for collecting epigenetic data from the environment, imagination in humans is more or less equivalent to flight in birds, assuming that freedom of mind is equal to freedom of movement, in principle.

References

S.T. Asma, *The Evolution of Imagination* (University of Chicago Press, Chicago, 2017)

F. Baluska, S. Mancuso, D. Volkmann, P.W. Barlow, The 'root-brain' hypothesis of Charles and Francis Darwin: Revival after more than 125 years. Plant Signal. Behav. **4**, 1121–1127 (2009)

D. Bohm, *Wholeness and the Implicate Order* (Routledge and Keegan Paul, London, 1980)

C. Brozzoli, A.C. Roy, L.H. Lidborg, M. Lövdén, Language as a tool: Motor proficiency using a tool predicts individual linguistic abilities. Front. Psychol. **10**, 1639 (2019)

D. Chalmers, Facing up to the problem of consciousness. J. Conscious. Stud. **2**, 200–219 (1995)

J. Henrich, *The WEIRDEST People in the World* (Farrar, Straus and Giroux, New York, 2020)

B.A. MacDonald, M. Sund, M.A. Grant, K.L. Pfaff, K. Holthaus, L.I. Zon, R. Kalluri, Zebrafish to humans: Evolution of the alpha3-chain of type IV collagen and emergence of the autoimmune epitopes associated with Goodpasture syndrome. Blood **107**, 1908–1915 (2006)

L.R. Machado, B. Ottolini, An evolutionary history of defensins: a role for copy number variation in maximizing host innate and adaptive immune responses. Front. Immunol. **6**, 1 (2015))

H.R. Maturana, F.J. Varela, *Autopoiesis and Cognition* (D. Reidel, Dordrecht, 1980)

T. Nagel, What is it like to be a bat? Philos. Rev. **83**, 435–450 (1974)

S. Piraino, F. Boero, B. Aeschbach, V. Schmid, Reversing the life cycle: Medusae transforming into polyps and cell transdifferentiation in Turritopsis nutricula (Cnidaria, Hydrozoa). Biol. Bull. **190**, 302–312 (1996)

L. Thompson, *Robert Frost: The Early Years, The Years of Triumph, and The Last Years* (Holt Rinehart, New York, 1966)

J.S. Torday, Homeostasis as the mechanism of evolution. Biology (Basel) **4**, 573–590 (2015a)

J.S. Torday, A central theory of biology. Med. Hypotheses **85**, 49–57 (2015b)

J.S. Torday, Consciousness, redux. Med. Hypotheses **140**, 109674 (2020)

J.S. Torday, W.B. Miller Jr., Terminal addition in a cellular world. Prog. Biophys. Mol. Biol. **135**, 1–10 (2018)

J.S. Torday, V.K. Rehan, *Evolutionary Biology, Cell-Cell Communication and Complex Disease* (Wiley, Hoboken, 2012)

S. Xu, P. Liu, Y. Chen, Y. Chen, W. Zhang, H. Zhao, Y. Cao, F. Wang, N. Jiang, S. Lin, B. Li, Z. Zhang, Z. Wei, Y. Fan, Y. Jin, L. He, R. Zhou, J.D. Dekker, H.O. Tucker, S.E. Fisher, Z. Yao, Q. Liu, X. Xia, X. Guo, Foxp2 regulates anatomical features that may be relevant for vocal behaviors and bipedal locomotion. Proc. Natl. Acad. Sci. U. S. A. **115**, 8799–8804 (2018)

W. Zheng, Z. Wang, J.E. Collins, R.M. Andrews, D. Stemple, Z. Gong, Comparative transcriptome analyses indicate molecular homology of zebrafish swimbladder and mammalian lung. PLoS One **6**, e24019 (2011)

Chapter 16
Cellular Evolution of Language

Introduction

Reasoning after the fact and making up "just-so stories" is much easier than actually determining the "how and why" of biology. But we are at a watershed moment in human history, resources becoming more limited, and huge pressures for food and medicine brought on by climate change in combination with globalization of economies. We need a credible, testable, and refutable narrative for human evolution so that we can rationally and predictably affect control of our lives.

Cell-Cell Communication as the Basis for Evolution

Deconvoluting the evolution of vertebrate physiology, acting as a "blueprint" for understanding the human condition based on cell-cell communication provides such a path (Torday and Rehan 2017. Evolution, the Logic of Biology. Wiley, Hoboken). In that vein, an effort has been made to trace the physiologic evolution of oral and written language, a problem that has been debated formally ever since the publication of Darwin's The Descent of Man, and Selection in Relation to Sex (1871). It should be borne in mind that to date every other attempt to decipher the evolution of language has been descriptive and synchronic (same space-time), whereas the only way to understand human physiologic traits, including language, is to transcend space-time diachronically (across space-time) in order to factor out the material "signature" that clouds our judgment in this regard. David Bohm characterizes this clouded perspective as the explicate order in his book Wholeness and the Implicate Order (1980).

Previously, it has been demonstrated that the evolution of the lung can be traced back to the unicellular state using functional genomic interrelationships between

J. S. Torday, *Hormones and Reality*, https://doi.org/10.1007/978-3-030-93691-4_16

lipids, gravity, and oxygen, mediated by cell-cell communications under the constraints of homeostasis. That approach culminated in recognition of the first principles of physiology (Torday and Rehan 2009). When that perspective is adopted, moving forward it predicts many physiologic traits that up until then had been seen as descriptive dogma rather than causally predictive explanations.

Using the same prima facie approach, language may also be reverse-engineered since it too is a physiologic trait. There are large gaps in our ability to trace the stepwise evolution of language, but that is a "chicken and egg" problem largely due to thinking about evolution from its ends, not its means, its consequences, not its etiology, so scientists have not been formulating experiments along these lines up until now. But that can be corrected by formulating hypotheses based on cell-cell communication mechanisms for both ontogeny and phylogeny, as has previously been demonstrated.

For example, the hormone leptin has the same developmental effect on the frog lung as it does on the mammalian lung (Torday et al. 2009). That is experimental evidence for evolution as a continuum involving epigenetic change rather than random genetic mutations, the principle of randomness obviating the option of seeing the process of lung evolution step-by-step as a systematic approach based on homeostatic control.

As a case in point, the debate over whether evolution is gradual (Darwin) or "punctuated" (Gould and Eldredge 1993) has remained an open question. It has been proposed that by focusing on the cellular-molecular components of physiologic evolution that the process is both gradual and punctuated depending upon the method you use. That is analogous with Neils Bohr's complementarity explanation for whether light is a particle or a wave (Bohr 1928). Indeed, the lowest-level origins of physiologic evolution for gas exchange – water-electrolyte balance in reference to language – remains cryptic, until the water-land transition some 500 million years ago, at which time the adaptation to land caused the overt phenotypic expression of such traits for the first time. Most of the terrestrially adaptive traits, like air breathing, micturition, skin barrier function, and the brain, were accounted for by the duplication of the parathyroid hormone-related protein receptor (Pinheiro et al. 2012); the terrestrially adaptive evolution of language is linked to the duplication of the β-adrenergic receptor during the water-land transition (Aris-Brosou et al. 2009), which facilitated lung evolution (Fig. 16.1), the latter mediating bipedalism, freeing the forelimbs for toolmaking and oral language, control of both localized to the area of Broca. Written language emerged as a consequence of the synergistic interactions between toolmaking and language.

This stepwise integration of physiology has been hypothesized to be the basis for consciousness, particularly because it explains David Chalmers' "hard problem," as well as Krishnamurti's "the observer is the observed." And these functionally integrated interrelationships would also provide a rationale for the capacity of language to effectively express ourselves not only in daily parlance but in literature, poetry, and the lyrics for musical composition alike, literally expressing our most visceral feelings and thoughts.

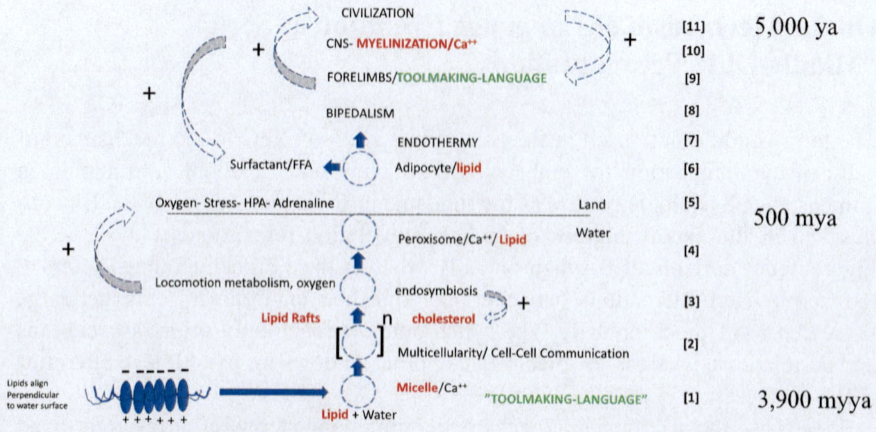

Fig. 16.1 On the evolution of language. Timeline in "years ago" (= ya) on the left-hand side. The formation of micelles (Aris-Brosou et al. 2009) led to multicellular organisms (represented by the "n" as an exponent) (Berwick and Chomsky 2017), the introduction of cholesterol into the cell membrane (Bohm 1980), the evolution of the peroxisome (Bohr 1928), the water-land transition and duplication of the β-adrenergic receptor (Carter 1974), the evolution of endothermy (Darwin 1871; De Duve 1969), leading to bipedalism (Gould and Eldredge 1993), freeing the forelimbs (Hawking 2011), selection pressure for myelinization of neurons to facilitate calcium flux (Kuhn 1996), and culminating in the evolution of civilization (Lawson et al. 1978)

Bottom-Up, Top-Down, Middle-Out

There are three modes of evolution – bottom-up, top-down, or middle-out. The first two are descriptive, lacking any mechanistic explanation. Darwinian evolution based on natural selection is top-down, but without any mechanistic explanation, it is merely an untestable metaphor. Top-down selection has also been referred to as "downward causation," as if it were a mechanism, but again it is teleologic. Contrary to that, the middle-out process is highly mechanistic, based on cell-cell communications mediated by soluble growth factors and their cognate receptors signaling for homeostasis. Beginning with embryologic development for structure and function, the process culminates in physiologic control. Seen in this light, injury-repair is part of this continuum. This cellular-molecular approach can be exploited to understand the process of evolution based on experimental evidence that is Popperian, i.e., testable and refutable. Furthermore, it is predictive, offering ways of solving such complex problems facing humanity as climate change and the practice of preventive medicine.

On the Mechanism of Language Evolution as Serial "Middle-Out" Preadaptations

The term "middle-out" used in the title of this section refers to the mechanism of cell-cell communication for embryologic structure and function, culminating in homeostatic physiologic control as the fundament for language (Fig. 16.1). By contrast, all of the extant theories of language evolution refer to steps 9–11 in the Figure, seen "horizontally," synchronically, whereas the current hypothesis refers to the step-by-step interactions between cells and their environment epigenetically, from step 1–11 diachronically. The synchronic approach only offers associations and correlations, whereas the diachronic approach allows for causal, testable/refutable hypotheses.

Please use the accompanying schematic, numbered stepwise on the right-hand side: (Aris-Brosou et al. 2009) lipids are amphiphilic, meaning that they have both a positively and negatively charged "pole." When immersed in water, lipid molecules self-assemble perpendicularly to the surface of the water, with their negative poles pointed downward into the water (hydrophilic) and their positively charged poles pointing upward into the air (hydrophobic). In this configuration, the lipids pack together tightly, reducing the surface tension of the water surrounding them, spontaneously leading to the formation of spheres with semipermeable membranes due to the force of gravity (Claassen 1996). Such lipid spheres are referred to as micelles, in effect forming the explicate order boundary with the implicate order. (Berwick and Chomsky 2017) Subsequently, competition between eukaryotic and prokaryotic unicellular organisms, the latter forming pseudo-multicellular states using biofilm and quorum sensing, led to true multicellular organisms formed by soluble growth factors signaling for structure and function, initiating life, as referred to in step (Aris-Brosou et al. 2009) based on the first principles of physiology – negentropy, chemiosmosis, and homeostasis; cholesterol synthesis (Bohm 1980) was made possible by the amount of oxygen in the atmosphere, one molecule of cholesterol requiring 11 atoms of oxygen to form. But the preadaptation for the synthesis of cholesterol in response to larger and larger amounts of oxygen was based on cellular cooperativity (Aris-Brosou et al. 2009) and the exploitation of lipids to form the first cell (Berwick and Chomsky 2017); subsequently, calcium levels in the primordial ocean gradually rose due to the formation of carbonic acid from carbon dioxide dissolved in water leaching calcium from the bedrock. To cope with the existential threat of calcium causing toxicity for lipids, cells sequestered calcium in their endoplasmic reticulum. But under oxidative stress, the calcium stores leaked out into the cytoplasm of the cell, referred to as endoplasmic stress, compensated for by the evolution of peroxisomes (De Duve 1969) (Bohr 1928), which utilize lipids to "buffer" the toxic effect of excess calcium in the cytoplasm, thus repeating steps (Aris-Brosou et al. 2009; Berwick and Chomsky 2017; Bohm 1980) in cellular evolution; (Carter 1974), the hypoxic stress caused by the stepwise evolution of the lung enlisted the previous four biologic traits, mediated by the increased production of adrenaline, causing increased secretion of lung surfactant by the alveoli (Lawson et al. 1978), facilitating increased oxygenation by acutely allowing greater expansion of the alveoli. A "side effect" (Darwin 1871) of hypoxic stress-induced adrenaline production was the breakdown of adipocytes, releasing free fatty acid into circulation, increasing metabolic activity, producing heat as the earliest form of endothermy (De Duve 1969); again, this trait was contingent on the prior six traits; endothermy (Gould and Eldredge 1993) facilitated bipedalism since standing on two legs requires more efficient metabolism for energy production than quadrupedalism does. Berwick and Chomsky's Why Only Us (2017) hypothesize a small growth factor change as the missing link from chimpanzees to humans; the roles of FoxP2 in development of the lung, gut, and bone as pleiotropy implicates Foxp2 in bone adaptation for bipedalism unique to humans, i.e., selection pressure for Foxp2 links bipedalism with language formation (Xu et al. 2018). Consequently, bipedalism freed the forelimbs for toolmaking (Hawking 2011), placing positive selection pressure on the myelinization of neurons in the central nervous system (Kuhn 1996), as a consequence of steps (Aris-Brosou et al. 2009; Berwick and Chomsky 2017; Bohm 1980; Bohr 1928; Carter 1974; Darwin 1871; De Duve 1969; Gould and Eldredge 1993; Hawking 2011), and because the hands were preoccupied with toolmaking, bipedalism also gave rise to language as a derivative of toolmaking (Lawson et al. 1978). Evidence for the coevolution of motor skills and language relies on the fact that both of these functions reside in the area of Broca – all great apes possess an area of Broca, but only humans are bipedal, generating the positive selection pressure for toolmaking and language. Further synergistically positive selection pressure on toolmaking and language led to further adaptation underpinned by steps (Aris-Brosou et al. 2009; Berwick and Chomsky 2017; Bohm 1980; Bohr 1928; Carter 1974; Darwin 1871; De Duve 1969; Gould and Eldredge 1993; Hawking 2011; Kuhn 1996), giving rise to civilization and culture, in the aggregate.

Language as a derivative of toolmaking forms the basis for all human endeavor – literature, music, art, social sciences, geology, economics, sciences, psychology, etc. As such, the mechanism for the evolution of language can form the prototype for deconvoluting the other disciplines, forming the basis for a periodic table of education. This would constitute a paradigm shift in the way we think about communicating knowledge. In his book The Structure of Scientific Revolutions (1996), Kuhn stated that the hallmark of a paradigm shift is a change in the language, providing a rationale for focusing on it. And since language is hypothesized to be derived from toolmaking, it helps to deconstruct other disciplines such as literature, for example, offering ways that lead to deeper understanding of the human condition through narrative; perhaps even closer to the core concept of toolmaking is poetry, which is more inferential than descriptive, alluding to the implicate versus the explicate order. Take, for example, Emily Dickinson's characterization of "hope as the thing with feathers," merging verse with biology as a unity, or Robert Frost, stating that "Life is that which can mix oil and water," or his poem "The Secret Sits:"

We dance round in a ring and suppose,

But the Secret sits in the middle and knows.

Or take, for example, the art as a way of abstracting ideas, like Henry Moore's sculptures with large holes through them (Fig. 16.2), asking us to decide whether it is the material portion of the sculpture or the negative space that is of primary significance or both, or economics as the materialization of human physiology, begging the question whether it is "supply and demand" or "free energy" that is more faithful to the discipline.

The key to understanding the seemingly directed evolution of language, as portrayed above, is that the formation of the first cell by lipids allowed for the formation of the first principles of physiology – negative entropy, chemiosmosis, and homeostasis. All of the subsequently evolved traits are in reference to the maintenance and perpetuation of those principles.

Fig. 16.2 Henry Moore sculpture. Note the hole or negative space

Self-Referential Self-Organizing Self-Authorship

Given that the vertical integration of cellular evolution is the process by which language has evolved raises the question of its origin. The initial step in the progression is the answer, because the immersion of lipids in water, aligning at the water-air interface, their negatively charged ends pointed downward into the water because they are hydrophilic, their positively charged ends pointing skyward is the beginning of the orientation of life toward its ambiguous origin (Torday and Miller 2017). And the tighter the lipids pack together, the more effectively they reduce the surface tension of the water, enabling the formation of primitive protocells in response to the force of gravity. In so doing, they form the explicate order that Bohm describes in his book Wholeness and the Implicate Order (1980), that subjective version of the totality we have exploited in order to survive.

Micelles float at the surface of the ocean, consequently being warmed by the sun by day, liquifying and deforming. In the absence of the sun, the micelles recapitulate their original form since they exhibit hysteresis or molecular memory. That iterative warming up and cooling down tended to trap more and more calcium ions within the micelles as they expanded and contracted day and night, ultimately presenting an existential problem since too much calcium is toxic for lipids, eventually causing them to denature. However, a subset of such micelles hypothetically formed calcium channels that mediated the uptake and release of calcium ions, facilitated by the concomitant uptake and release of water molecules due to osmosis. That rhythmic uptake and release of calcium and water would have generated the pulsatility characteristic of both unicellular organisms and multicellular organisms alike. Importantly, the dynamic hysteretic interrelationship between lipids and calcium was hypothetically the origin of the molecular memory that is existential for evolution, the organism needing to remember its historic encounters by which it copes in order to combat future life or death encounters.

A prototypical example of such molecular memory is the advent of cholesterol synthesis, which Konrad Bloch, who discovered the biosynthetic pathway for cholesterol, referred to as a “molecular fossil” (Miao et al. 2002). He reasoned that since it takes 11 atoms of oxygen to produce one molecule of cholesterol, there had to have been enough oxygen in the atmosphere to do so. However, that was reasoning after the fact, which we know is illogical. Based on the serial preadaptation approach, the advent of cholesterol referenced the exploitation of the lipid molecules used to form life in the first place.

The insertion of cholesterol into the cell membrane fundamentally advanced eukaryotic evolution, thinning the phospholipid bilayer, increasing oxygen uptake, metabolism, and locomotion, the latter being due to increased cytoplasmic streaming. The presence of cholesterol in the cell membrane also formed the basis for the lipid rafts from which cell-surface receptors generated intercellular signaling to form multicellular organisms. These properties predicted many future vertebrate traits.

It was during the water-land transition some 500 million years ago that such lipid-related traits again proved existential. The β-adrenergic receptor duplicated (Aris-Brosou et al. 2009), amplifying that signaling pathway throughout the organism. Because the lung was evolving from the swim bladder in the effort by boney fish to adapt to land, there were episodes of physiologic stress due to the inadequacy of the nascent lung to provide enough oxygen, hypoxia being the most potent stressor known in vertebrate physiology. In response, the hypothalamic-pituitary-adrenal axis was activated, stimulating secretion of adrenocorticotrophic hormone (ACTH), which stimulated adrenocortical production of cortisol, which in turn stimulated the production of adrenaline by the adrenal medulla. Adrenaline stimulated the secretion of surfactant by the alveoli (Lawson et al. 1978), rendering them more distensible, transiently increasing the amount of oxygen in circulation, acutely relieving the constraint on lung evolution; long-term, the over-distension of the alveoli caused increased secretion of parathyroid hormone-related protein, which, in turn generates more alveoli. As a side effect, adrenaline caused the disruption of fat cells, releasing free fatty acid into the circulation, increasing metabolism and thus body heat. That was the ad hoc origin of warm-bloodedness or endothermy.

The advent of endothermy allowed for bipedalism or walking on two legs, since bipedalism requires much more energy than walking on four legs. And given the preadaptive reference to earlier traits, the upright posture may have been referencing the perpendicular alignment of lipid molecules in water at the inception of life, particularly given the fundamental role of bipedalism in freeing the forelimbs for toolmaking. As an intermediate step in the progression for bipedalism, eukaryotes evolved deuterostomy or the process of development from the anus to the mouth, causing selection pressure for terminal addition, underpinned by vagal evolution.

The production of tools may have been the origin of language from a practical standpoint since the hands were preoccupied with toolmaking, obviating manual gesticulations for communication. The forming of an arrowhead from a piece of flint is conceptually not very different from conceiving a thought and verbalizing it by generating a subject, verb, and object in a sentence (Wittgenstein). Positive selection pressure for the combined action of hands and language for toolmaking has promoted myelinization of neurons in the brain, progressively enhancing mental capacity. That synergy for toolmaking, language, and the central nervous system fostered written language, giving rise to social systems and civilization.

Holism

Ontologically, the Big Bang provided a point source for the origin of the universe, first detected as the microwave background that echoed that explosive event (Penzias and Wilson 1965). A homologous point source was recently identified for the unicellular origin of life (Torday and Miller 2016); the continued existence and evolution of eukaryotes was ensured by the biosynthesis and insertion of cholesterol into the cell membrane, facilitating vertebrate metabolism, locomotion, and respiration, the three foundational traits of vertebrate evolution. As a result, the cell membrane became more compliant, facilitating endocytosis and exocytosis, allowing for freer access of environmental factors to the cytoplasm by substances that would otherwise have killed eukaryotes off long ago such as heavy metals, ions, and gases, instead compartmentalizing such substances with endomembranes, making them useful as physiologic functions.

Seeing the cell perpetually internalizing the environment as a continuum for niche construction provides a way for conceiving of Gaia, James Lovelock's "Earth Mother." In this way of thinking from the unicell forward, biology being molded by the ever-changing environment, over and over again, the continuum of physics and biology becomes self-evident.

Organisms cope with environmental existential threats by repurposing genetic traits previously used successfully over the course of their evolution. For example, the advent of cholesterol in the cell membrane was characteristic of the overall process of vertebrate evolution beginning with the implementation of lipids that originated in pulsars in deep space. Serving homeostasis with reference to the first principles of physiology, lipids were utilized as antioxidants, lipid rafts for ligand-receptor cell-cell signaling, and as substrate for the steroid hormones. This is referred to as symbiogenesis, which is applicable to all eukaryotes, from protozoa to humans and every organism in between. Internalizing the environment, organisms have adapted to it, evolving internal organs over the course of vertebrate evolution. In tandem, mechanisms for detecting change in the environment ranging from the unicellular cell membrane to evolved organs of olfaction have been vertically integrated into the nervous systems of more complicated organisms. Taken as a whole, these iterative processes are our consciousness, or mind, as the way in which we intuit our environment. This way of understanding consciousness is radically different from conventional ways of thinking about consciousness as being in our heads or extending into the environment.

Cell-Cell Communication as "Parts-to-Wholes" Language Evolution

The laws of physics govern all of the natural laws, begging the question as to how physics determines biology. Physicists like Prigogine and Steggers (1984) and the polymath Polanyi (1968) attempted to answer this question and failed, concluding that biology is too complicated. However, that may be due to looking at the problem from its ends instead of its means; in contrast to that, there commonalities between physics and biology emanating from their shared origin in the singularity, thought to have given rise to the Big Bang (2011). In The Life of the Cosmos (1999), Lee Smolin has hypothesizes that the "evolution" of both stars and black holes can be explained based on Darwinian evolution theory. It is feasible that in the wake of the Big Bang, an "equal and opposite reaction" formed matter due to homeostatic self-referential self-organization. Chemistry and biology may have formed on that basis, though we may not be able to know the origins of the totality since we were not there at the time. Yet, we can surmise how and why it occurred based on scientifically testable and refutable hypotheses. It has been speculated that since lipids were critically important in the evolution of eukaryotes, they played a vital role in the origins and initial conditions of life, evolution being exaptive or co-optional. When lipids were submerged in water they may have formed the basis for life since they were components of the frozen snowball-like asteroids that formed the earth's ocean. Physicochemically, lipids spontaneously form micelles in water, exhibiting hysteresis as molecular "memory," able to recall their shape and size despite being deformed and reformed iteratively, which is necessary for evolution – the lipid membranes that delineated the internal and external environments of the cell generated life as an ambiguity.

Under the above circumstances, several aspects of quantum mechanics apply, such as the Pauli exclusion principle, non-localization, the Heisenberg uncertainty principle, coherence, quantum entanglement, and non-localization. Regarding the Pauli exclusion principle, the first three quantum variables are deterministic, but the fourth is probabilistic (Morowitz 2004). Consequently, life exists between the boundaries of determinism and free will, and biology complies with physics as the coexistence of both determinism and free will. Biology mimics the quantum mechanical principles based on homologies with the first principles of physiology. Similarly, the pleiotropic distribution of genes throughout the body accounts for the "spooky" (Yin et al. 2013) quantum mechanical principle of non-localization. In his book What is Life? Schrodinger (1944) related that the free energy or entropy of the cell is negative, in contrast to the positive entropy outside of the cell, so life began as an ambiguity; that ambiguity, or uncertainty, is the driver for the subsequent evolution of the organism over time, both developmentally and phylogenetically. For example, Niels Bohr explained the seeming duality of light as both particle and wave, as what he called complementarity, which refers to this artifact being due to differences between the ways that light is observed. That perspective is homologous with the ambiguity of the cell, coping with such paradoxes, as described above. Both physics and life exist in limbo due to their common dependence on "uncertainty," which allows us to contend with it, even at the quantum mechanical level – *like dissolves like*.

Biology needs the predictive value of quantum mechanics to become a "hard" science as it relates to evolution; the following is a means for understanding biology mechanistically, allowing it to effectively interface with quantum mechanics, revealing the common source of physics and biology as the singularity/Big Bang. By internalizing the environment through symbiogenesis, the unicell became the first niche construction. The ensuing communication between cells in support of metabolic cooperativity was realized through niche construction, in combination with the inheritance of epigenetic marks, giving rise to larger and larger organismal communities. The systematic dissemination of these collective properties became Gaia, the concept of the organic unity of the earth. Fluctuations in the levels of oxygen and carbon dioxide in the earth's atmosphere have caused physiologic stresses on life forms over the course of the earth's history. Based on that premise, the evolution of the mammalian lung was traced from its unicellular state forward by aligning the phenotypic changes in the gas exchanger with the increases and decreases in atmospheric oxygen; with that perspective, evolution could then literally be viewed as a homology with quantum mechanics (QM). Reducing life to its elements, expressed as the first principles of physiology enables seeing the one-to-one relationships between QM and physiology. That is comparable with Mendeleev's periodic table, reducing the hierarchy of the elements to the number of protons in their nuclei. Consequently, QM can be applied directly to evolution as a homology with the Pauli exclusion principle, non-localization, and Heisenberg uncertainty principle. For example, Pauli exclusion principle stipulates that no two electrons can have the same spin in an atom. The spin of an electron is determined by four quanta of energy, the first quantum determining the quantum of the second and third; the last quantum number is time-based, rendering it probabilistic. Therefore, each electron exists between the boundaries of determinism and probability. Similarly, the cell resides between negentropy and chemiosmosis, which are deterministic, and the free will of homeostasis, which is probabilistic. Nonlocal distribution of physical properties throughout the totality of the cosmos is another homology with the cell. Non-localization produces pleiotropic traits by acquiring the same gene over the course of its history under different environmental conditions.

Cells as Fractals of Physiology

The cell may be seen as a fractal of physiology and beyond that as a fractal of the cosmos. Seen in this holistic way, language "verbalizes" the vertical integration of physiology from the unicell to consciousness. This way of thinking helps understanding of such phenomena as near-death experiences, out-of-body experiences, and Maslow peak experiences as natural consequences of the synchrony of cell-cell communications, resulting in unified "fields" of calcium waves. These experiences allow the convergence of individual consciousness with that of the cosmos as the totality of natural laws.

Physiology: Top-Down, Bottom-Up, or Middle-Out

The question of whether physiology is top-down or bottom-up has been debated for decades. The question is what the basis for such control systems is. The middle-out approach for understanding the evolution of language begins with structure and function being determined by cell-cell communication, mediated by soluble growth factor-receptor signaling that begins with the zygote and persists until death. It originates with the unicellular state and, beyond that, to the singularity. And because of its validity, it provides a number of explanations for dogma – the nature of the cell, function of homeostasis, pleiotropy, life cycle, and heterochrony – all of which are intractable based on descriptive biology.

Discussion

Linguistics has a long history, dating back to Darwin's The Descent of Man (Darwin 1871), speculating that "the survival of certain favored words in the struggle for existence is natural selection." From then on, theories of language have only been framed in its overtly synchronic forms, reasoning after the fact. In contrast to that, the present hypothesis that language evolved as the process of cell-cell communication, which determines embryogenesis, physiologic homeostasis, and injury-repair, transcends space-time diachronically. In so doing, it eliminates the stigma of anthropocentric thinking. By focusing on language as a natural consequence of a hierarchically derived, holistic integration of the cell with its environment allows for a novel, a priori way of understanding the true origin and nature of linguistics that is testable and refutable rather than being subjectively inductive and speculative.

A systematic understanding of linguistics is critically important as we move into the Anthropocene, a man-made world, accepting artificial intelligence without mechanistically predictive knowledge of our evolution. In effect, we are tampering with aspects of human traits that we do not understand either ontologically or epistemologically, potentially exposing ourselves to the hazards of our ignorance. The classic example is the anthropic principle (Carter 1974) that we have been put here and have had the great good fortune to end up in a place that accommodates our physiology. Actually, we are not "in" this place; we are "of" this place, literally. Carl Sagan used to close his cosmos TV show by saying "The cosmos is within us. We are made of star-stuff. We are a way for the universe to know itself." He intuited this without experimental evidence but was probably influenced by his wife, Lynn Margulis Sagan, who promoted endosymbiosis theory, predicated on the idea that we have evolved by internalizing things in our environment that would otherwise have destroyed us. So she was providing the data for Sagan's trope. The discovery of soluble growth factors and their receptors formed the basis for embryogenesis and physiology to understand the underlying mechanism.

The value of the above perspective is most evident in the deconvolution and recapitulation of the evolution of language. Language emerges from the depths of the cosmos, forming a logic that can be seen in the periodic table of elements, the elements produced by the stars as they generated light from matter, beginning with the lightest, hydrogen, and ending with the heaviest elements. The cosmos is 13.8 billion years old, and it was 9 billion years until the earth was formed. The first cell appeared once the ocean formed from the frozen water delivered by the snowball-like asteroids pelting the planet. The polycyclic hydrocarbons that were also present spontaneously formed micelles or prototypical "cells." The resulting delineation of "inside" and "outside" distinguished Bohm's explicate order from the implicate order for the first time. That was the origin of language as a "dialogue" between life and nonlife, which culminated in oral and written language, the latter being the key to the rise of civilization. It is imperative that we understand this intimate relationship between our physiology and our environment through language. We must extricate ourselves from the deception that Robert Trivers has determined has thus far been our undoing (Trivers 2014). To do otherwise puts us at risk for extinction.

References

S. Aris-Brosou, X. Chen, S.F. Perry, T.W. Moon, Timing of the functional diversification of alpha- and beta-adrenoceptors in fish and other vertebrates. Ann. N. Y. Acad. Sci. **1163**, 343–347 (2009)

R.C. Berwick, N. Chomsky, *Why only us?* (MIT Press, Cambridge, 2017)

D. Bohm, *Wholeness and the implicate order* (Routledge and Kegan Paul, London, 1980)

N. Bohr, The quantum postulate and the recent development of atomic theory. Nature **121**, 580–590 (1928)

B. Carter, *Large number coincidences and the anthropic principle in cosmology. IAU symposium 63: confrontation of cosmological theories with observational data* (Reidel, Dordrecht, 1974)

D.E. Claassen, B.S. Spooner, Liposome formation in microgravity. Adv. Space Res. **17**, 151–160 (1996)

C. Darwin, *The descent of man, and selection in relation to sex* (John Murray, London, 1871)

C. De Duve, Evolution of the peroxisome. Ann. N. Y. Acad. Sci. **168**, 369–381 (1969)

S. Gould, N. Eldredge, Punctuated equilibrium comes of age. Nature **366**, 223–227 (1993)

S. Hawking, *A brief history of time* (Bantam, New York, 2011)

T.S. Kuhn, *The structure of scientific revolutions* (University of Chicago Press, Chicago, 1996)

E.E. Lawson, E.R. Brown, J.S. Torday, D.L. Madansky, H.W. Taeusch Jr., The effect of epinephrine on tracheal fluid flow and surfactant efflux in fetal sheep. Am. Rev. Respir. Dis. **118**, 1023–1026 (1978)

L. Miao, M. Nielsen, J. Thewalt, J.H. Ipsen, M. Bloom, M.J. Zuckermann, O.G. Mouritsen, From lanosterol to cholesterol: Structural evolution and differential effects on lipid bilayers. Biophys. J. **82**, 1429–1444 (2002)

H. Morowitz, *The emergence of everything* (Oxford University Press, Oxford, 2004)

A.A. Penzias, R.W. Wilson, A measurement of excess antenna temperature at 4080 mc/s. Astrophys. J. **142**, 419–421 (1965)

P.L. Pinheiro, J.C. Cardoso, D.M. Power, A.V. Canário, Functional characterization and evolution of PTH/PTHrP receptors: Insights from the chicken. BMC Evol. Biol. **6**, 110 (2012)

M. Polanyi, Life's irreducible structure. Live mechanisms and information in DNA are boundary conditions with a sequence of boundaries above them. Science **160**, 1308–1312 (1968)

I. Prigogine, I. Steggers, *Order out of Chaos* (Bantam, New York, 1984)

E. Schrodinger, *What Is life?* (Cambridge University Press, Cambridge, 1944)

L. Smolin, *The life of the cosmos* (Oxford University Press, Oxford, 1999)

J.S. Torday, W.B. Miller, The unicellular state as a point source in a quantum biological system. Biology (Basel) **5**, 25 (2016)

J.S. Torday, K. Ihida-Stansbury, V.K. Rehan, Leptin stimulates Xenopus lung development: evolution in a dish. Evol. Dev. **11**, 219–224 (2009)

J.S. Torday, V.K. Rehan, *Evolution, the Logic of Biology* (Wiley, Hoboken, 2017)

J.S. Torday, W.B. Miller Jr., The resolution of ambiguity as the basis for life: A cellular bridge between Western reductionism and Eastern holism. Prog. Biophys. Mol. Biol. **131**, 288–297 (2017)

J.S. Torday, V.K. Rehan, Lung evolution as a cipher for physiology. Physiol. Genomics **38**, 1–6 (2009)

R. Trivers, *The folly of fools* (Basic Books, New York, 2014)

S. Xu, P. Liu, Y. Chen, Y. Chen, W. Zhang, H. Zhao, Y. Cao, F. Wang, N. Jiang, S. Lin, B. Li, Z. Zhang, Z. Wei, Y. Fan, Y. Jin, L. He, R. Zhou, J.D. Dekker, H.O. Tucker, S.E. Fisher, Z. Yao, Q. Liu, X. Xia, X. Guo, Foxp2 regulates anatomical features that may be relevant for vocal behaviors and bipedal locomotion. Proc. Natl. Acad. Sci. U. S. A. **115**, 8799–8804 (2018)

J. Yin, Y. Cao, H.-L. Yong, J.-G. Ren, H. Liang, S.-K. Liao, F. Zhou, C. Liu, Y.-P. Wu, G.-S. Pan, Q. Zhang, C.-Z. Peng, J.-W. Pan, Bounding the speed of 'spooky action at a distance. Phys Rev Lett **110**, 260407 (2013)

Chapter 17
Neoteny and Human Evolution

Humans only differ from chimpanzees by about 1% of their DNA, yet there is a world of difference between us – characterized by physiologic features of a humans that resemble late-stage fetal chimpanzees. These fetal characteristics include the hair on the head, globular skull, ear shape, vertical plane face, absence of penile bone (baculum) in fetal male chimpanzees, vagina pointing forward in fetal apes, presence of a hymen in neonatal apes, and structure of the foot.

How then did we evolve from our primate cousins? To date, there are many observations that are consistent with humans as neotenous apes, but the cause of the phenomenon has not been explained. The following is a hypothetical mechanism that is primarily consistent with our oversized heads and brains. A previous publication had made the case for the role of warm-bloodedness allowing for walking on two legs, given that to do so required more energy than crawling on all fours. Endothermy provides more energy than does poikilothermy or cold-bloodedness. That concatenation of traits, in turn, evolved from the three documented receptor gene duplications that occurred during the water-land transition – parathyroid hormone-related protein receptor (PTHrPR) (Pinheiro et al. 2012), glucocorticoid receptor (GR) (Bridgham et al. 2006), and ß-adrenergic receptor (ßAR) (Aris-Brosou et al. 2009). All three of these receptor gene duplications amplified their specific physiologic properties for land adaptation, particularly in response to the stress of hypoxia, which facilitated the evolution of the lung. In that context, it is noteworthy that the levels of oxygen in the atmosphere over the course of the last 500 million years (the Phanerozoic) were marked by substantial increases and decreases, ranging from 15% to 30% (Berner 1999). Much has been said about the increases in oxygen causing giantism, but literally nothing has been written about the concomitant declines, despite the fact that hypoxia is the most potent stimulus for the hypothalamic-pituitary-adrenal (HPA) axis. During intermittent periods of hypoxia in the transition from water to land, the stimulation of the HPA caused production of adrenalin by the adrenal gland, relieving the shortage of circulating oxygen by stimulating lung surfactant production by the alveoli (Lawson et al.

J. S. Torday, *Hormones and Reality*, https://doi.org/10.1007/978-3-030-93691-4_17

1978), making them more distensible, further expanding to increase oxygen uptake. Over the course of vertebrate evolution on land, the overdistension of the alveoli led to increasing numbers of alveoli since the stretching of the alveolar wall stimulates the production of PTHrP by the alveolar type II cells, which increases the number of alveoli.

As a side effect, adrenalin also stimulates the release of fatty acids from fat cells, causing increased body temperature since fatty acids are a highly efficient source of metabolic energy. Over time, that adaptive increase in body temperature became constitutive through the genetic role of oxytocin for thermoregulation (Kasahara et al. 2007).

Evolutionarily, bipedalism freed the forelimbs for specialized function in the form of toolmaking, aided by the prehensile thumb. And language as a form of toolmaking put further positive selection pressure on our highly evolved central nervous system (brain). We are born with only about 30% of our adult brain capacity because as the brain evolved under the influence of positive selection pressure, it got larger and larger, aiding and abetting toolmaking and language, expanding our consciousness as humans. The evolving head eventually became too large to pass through the birth canal, putting selection pressure on being born prematurely, which is classically a fixed value of about 10% (nowadays it is 13% due to the rise in the number of multiple births as a consequence of artificial reproduction) (Blencowe et al. 2012).

Prematurity is conventionally seen as a pathologic condition, yet it is a consequence of the evolution of the brain to the relative detriment of other organs. As a result, we experience a protracted childhood during which the brain must further develop to full capacity, literally extending until our early to mid- to late-twenties, though we know that the brain remains plastic throughout life, perhaps extending our juvenile behavior throughout our entire lives.

There is behaviorally based endocrine evidence for this in the phenomenon of subjective age, the sense that we feel younger than our chronological age in late life (Kwak et al. 2018). This is associated with the decline in adrena dehydroepiandrosterone production (Samaras et al. 2013), providing a physiologic basis for juvenile behavior, which projects all the way back to the same hormonal mechanism during teen-hood, whereas in that stage we tend to feel older than our chronological age. Narcissism and risk-taking behavior are typical of this state of being, as well as inventiveness and promiscuity – all highly characteristic of our current human society – referred to as "consumer neoteny" (Oliver 2016). Such behavior is exacerbated by computer technology, which is designed to tap into our personal characteristics, feeding into our innate narcissism.

The above is predicated on the physical interactions between life forms and their environments, beginning with lipids immersed in water orienting vertically to the surface because they are amphiphiles, having both a positively and negatively charged end, the negatively charged end being hydrophilic, so it is attracted to the water, and the positively charged end pointing upward, away from the water. When such lipid molecules pack together, they reduce the surface tension of the surrounding water by disrupting the cohesive effect of the Vander Waals forces, allowing for the formation of the first protocell.

The primordial ocean that covered the earth was formed by ice-containing asteroids pelting the atmosphere-less planet; lipids formed by pulsars were also contained within those asteroids. When lipids are mixed with water, they will spontaneously produce micelles, small spheres formed by semipermeable lipid membranes. The interior of such protocells was a protected space for negative entropy, maintained by chemiosmosis and regulated by homeostasis, which have been termed the first principles of physiology. Those traits emerged from particles entering such protocells, gravity striking the curved surface of those micelles generating the energy for quantum entanglement under the influence of nonlocal events occurring in the cosmos. That sequence of events was the archetype for symbiogenesis, the process by which factors in the environment that posed an existential life-endangering threat were endogenized and compartmentalized, forming physiologic traits.

Such primitive cells deformed due to liquification when the sun warmed them as they floated on the surface of the ocean, and they reformed at night due to hysteresis when it got cold or molecular memory. The recursive expansion and contraction of those micelles caused trapping of calcium ions within them; high calcium levels would have posed an existential threat since calcium denatures lipids. However, those micelles that were able to form calcium channels for the control of entry and exit of calcium into and out of those micelles are our forebears. Memory is necessary for evolution to occur, and this conditional property was eventually replaced by genetically controlled nucleotides, i.e., RNA and DNA.

Over the course of evolution, there were various processes that "whipsawed" the human form, beginning with the gyrations of embryos as they grow and differentiate in utero, from the zygote to the morula and gastrula. Wolpert has said that gastrulation is the most important thing we will do over the course of our lifetime – that is the stage in embryologic development when the mesoderm is introduced in between the endoderm and ectoderm, conferring plasticity (West-Eberhard 2003).

Subsequently, our forebears evolved as deuterostomes, forming from the anus to the mouth. The evolution of bipedalism further reinforced that configuration, pointing upward away from the center of gravity in the earth. This orientation was reinforced structurally and functionally by the vagal nerve, the major nerve of the autonomic nervous system, which forms from the adrenals to the gut to the heart and head, terminating in the 12 cranial nerves. The 11th cranial nerve passes through the fossa at the base of the skull and enervates the tongue, facilitating speech. The other cranial nerves control facial gestures, which in turn affect the evolution of the brain. It is speculated that these gyrations are the basis for lateralization or left-right brain function. Lateralization refers to the left and right cerebral hemispheres, which control the right and left sides of the body, respectively. That "crossover" is reminiscent of what is described above for the evolution of form and function, referring all the way back to those lipids standing up vertically to the surface of the ocean. When we contemplate something in the environment it is with both the left and right hemispheres, resulting in discernment of the quantum mechanical state.

References

S. Aris-Brosou, X. Chen, S.F. Perry, T.W. Moon, Timing of the functional diversification of alpha- and beta-adrenoceptors in fish and other vertebrates. Ann. N. Y. Acad. Sci. **1163**, 343–347 (2009)

R.A. Berner, Atmospheric oxygen over Phanerozoic time. Proc. Natl. Acad. Sci. U. S. A. **96**, 10955–10957 (1999)

H. Blencowe, S. Cousens, M. Oestergaard, D. Chou, A.B. Moller, R. Narwal, A. Adler, C.V. Garcia, S. Rohde, L. Say, J.E. Lawn, National, regional and worldwide estimates of preterm birth. Lancet **379**, 2162–2172 (2012)

J.T. Bridgham, S.M. Carroll, J.W. Thornton, Evolution of hormone-receptor complexity by molecular exploitation. Science **312**, 97–101 (2006)

Y. Kasahara, Y. Takayanagi, T. Kawada, K. Itoi, K. Nishimori, Impaired thermoregulatory ability of oxytocin-deficient mice during cold-exposure. Biosci. Biotechnol. Biochem. **71**, 3122–3126 (2007)

S. Kwak, H. Kim, J. Chey, Y. Youm, Feeling how old I am: Subjective age is associated with estimated brain age. Front. Aging Neurosci. **10**, 168 (2018)

E.E. Lawson, E.R. Brown, J.S. Torday, D.L. Madansky, H.W. Taeusch Jr., The effect of epinephrine on tracheal fluid flow and surfactant efflux in fetal sheep. Am. Rev. Respir. Dis. **118**, 1023–1026 (1978)

M.A. Oliver, Consumer Neoteny: An evolutionary perspective on childlike behavior in consumer society. Evo Psychology **14**, 1–11 (2016)

P.L. Pinheiro, J.C. Cardoso, D.M. Power, A.V. Canário, Functional characterization and evolution of PTH/PTHrP receptors: Insights from the chicken. BMC Evol. Biol. **12**, 110 (2012)

N. Samaras, D. Samaras, E. Frangos, A. Forster, J. Philippe, A review of age-related dehydroepiandrosterone decline and its association with well-known geriatric syndromes: Is treatment beneficial? Rejuvenation Res. **16**, 285–294 (2013)

M.J. West-Eberhard, *Developmental plasticity and evolution* (Oxford University Press, Oxford, 2003)

Chapter 18
Life Is a Mobius Strip

Introduction

The mathematician Louis Kauffman informs us that if you cut a Mobius strip in half its edges are a Trefoil knot, which is reducible to a circle because it is a true mathematical knot. The cell is homologous with a mathematical knot since it must also be able to untie itself to form the egg or sperm in order for the organism to reproduce.

The relationship of the knot and the cell is thought-provoking from a biologic standpoint because the Trefoil knot is a homologue of the endoderm, ectoderm, and mesoderm, the three germ layers that interact to form the embryo, beginning with the zygote.

The Cell Membrane as the Prototype for the Mobius Strip

The cell membrane is like a Mobius strip because it forms a continuous topologic surface between the outer and inner environments of the cell, having generated the "inside" and "outside" when micelles first formed. When lipids were first submerged in water, they produced the "concept" of a Mobius strip because prior to that there was no inside or outside. The implicate order was one infinite plane. To understand this, let us go back to the origin of the cell as amphiphilic lipid molecules floating on the surface of the waters that covered the earth 100 million years after its formation. The negatively charged ends of the lipids pointed downward into the water, whereas the positively charged ends pointed upward. Once enough lipid molecules aligned with one another and packed together, they reduced the surface tension of the surrounding water by inhibiting the Van Der Waals forces that create the surface tension of water. In so doing, the lipids spontaneously form micelles or lipid spheres. Those micelles formed the explicate order as distinct from the implicate

J. S. Torday, *Hormones and Reality*, https://doi.org/10.1007/978-3-030-93691-4_18

order, as described in David Bohm's classic book Wholeness and the Implicate Order (Bohm 1980). The preadaptation for the micelles were the lipid molecules immersed in water, but there was no inside or outside; inside and outside only came into being once the micelles emerged from those lipid molecules. Therefore the micelle was the origin of the Mobius strip.

The hypothesis being tested is that the lipid barrier of the explicate order "conceived" the Mobius strip by dividing the implicate order into "inside and outside," formulating the concept of life in the process. That was the origin of consciousness, based on the first principles of physiology, referencing the singularity that predated the Big Bang (Hawking 2011).

Calcium Ion Fluxes, Micelles, and Semipermeable Membranes

Micelle semipermeable membranes allow particles to enter and exit, forming the basis for the protocell. Calcium released from the bedrock was prevalent in the primordial ocean. As carbon dioxide produced by plants built up in the atmosphere, it dissolved in the water to produce carbonic acid, hastening the dissolution of the bedrock, releasing ever-more calcium into the water (Case et al. 2007).

The accumulation of calcium ions within protocells sped up due to the sun warming them by day, causing the lipid membranes to liquify and enlarge; by night, the micelles cooled and returned to their original form due to hysteresis or "molecular memory." That repeated expansion and contraction resulted in the accretion of calcium ions within the protocells. Since calcium ions can denature lipids, a subset of protocells hypothetically evolved the capacity to control the entry and exit of calcium ions using calcium channels.

Calcium ion fluxes through the cell constitute the biologic flow of energy. When lung and bone cells are experimentally exposed to zero gravity, they lose their capacity to generate a calcium ion flow (Purevdorj-Gage et al. 2006). This indicates the critical role of gravity in the formation and initiation of life, and it even applies to abiotic micelles (Claassen and Spooner 1996). This causal relationship between the force of gravity and the "life force" reflects the integral relationship between physics and biology. It transcends the cell because Einstein's gravitational field theory states that when gravity impinges on a curved surface, it produces energy. Quantum entanglement would have entrained particles within the micelle, stabilized by such gravity-induced energy; the configuration of such particles would have referenced nonlocal events in the cosmos based on quantum mechanical principles. This foundational set of principles for the micelle would have provided for the emergence of symbiogenesis – the acquisition of factors in the environment that posed an existential threat during the "history" of the organism or what we commonly think of as cellular evolution.

Evolution of Multicellular Organisms

There was a juncture over the course of cellular evolution at which prokaryotes (bacteria) developed means for imitating multicellularity – namely, quorum sensing and biofilm – gaining a biologic advantage over unicellular eukaryotes. To be able to compete, eukaryotes evolved ways of effecting true multicellularity using cell-cell communications based on soluble growth factors signaling to their cognate receptors on neighboring cells. Starting from the fertilized egg, the animal and vegetal poles communicate using fibroblast growth factor to initiate cell division and differentiation that ultimately terminate in the offspring (Slack 1994). After birth the same cell-cell signaling mechanisms are implemented to sustain homeostasis within and between cells (Demayo et al. 2002). Breakdown in cell-cell communication results in remodeling of structure and function constrained by homeostasis. Physiologic stress can disrupt homeostatic control, causing production of reactive oxygen species (ROS). ROS can cause gene mutations and duplications (Storr et al. 2013), providing a mechanistic basis for "how and why" variation in cell-cell signaling occurs. By reestablishing homeostasis, a new set point is achieved and the cells involved will maintain that state as what we call evolution.

By example, the lung evolved from the swim bladder of boney fish, both being "gas exchangers." The swim bladder utilizes gases to feed efficiently at different water depths, inflating or deflating the bladder to control of buoyancy, facilitating metabolism and minimizing energy output. In the case of the lung, it mediates the uptake of oxygen for metabolism by the tissues and organs of the body, same genes, different purposes, which is referred to biologically as pleiotropy (Torday 2018).

The above description for the emergence of vertebrates from water to land exemplifies the big picture strategy of evolution. CO_2 built up in the atmosphere as plant life on earth evolved, causing a so-called "greenhouse effect," warming the ambient temperature of the atmosphere (Romer 1949). That process had a twofold effect on life, both reducing the amount of oxygen in the water and partially drying up the waters covering the earth, thus exposing land masses for habitation. The combination of these effects caused boney fish to move onto land. It was specifically the physostomous boney fish that were the basis for land vertebrates, with a pneumatic duct connecting the esophagus to the swim bladder, which is the structural homologue of the trachea in land vertebrates. And furthermore, there are cellular-molecular homologies between the pneumatic duct and the trachea that link them evolutionarily such as the expression of fibroblast growth factor 10 in the smooth muscle of both structures (Korzh et al. 2011).

Tiktaalik is the fossilized "history" for the water-land transition, representing the transition from fins to legs (Shubin et al. 2006). Such fossil evidence for evolution of the internal organs is impossible to ascertain since there is no such "hard" evidence, but there are cellular and molecular data that have been retained in the ontogeny and phylogeny of tissues and organs to document that history of the organism in the stepwise nature of cell-cell signaling, allowing recall of such events.

The fossilized skeletal evidence reveals that vertebrates attempted to live on land at least five separate times (Clack 2012), indicating a stepwise mechanism for evolving from the swim bladder to the lung. Lung surfactant constitutes the final common biomolecular pathway for this series of events, the soap-like material that prevents the alveoli from collapsing due to the prevailing hydrostatic force generated by the thin water layer lining the alveoli. Surfactant is produced by the swim bladder (Prem et al. 2000), preventing the walls of the bladder chamber from sticking together. The bladder chamber evolved into alveoli by progressively decreasing the size of the gas-exchange "sphere," effectively increasing the surface area-to-blood volume ratio for oxygenation (Orgeig et al. 2007). But for that to have occurred, tandem increases in surfactant activity were necessary since surface tension is reciprocally related to the diameter of a sphere (by the Law of Laplace). Consistent with that, as the lung evolved, both the amount and functional quality of lung surfactant increased commensurately (Lang et al. 2005), validating the predicted interrelationships. Over the course of land adaptation, other physiologic traits also evolved, particularly the endocrine system (Kleine and Rossmanith 2016). As fish emerged from water to land, from the swim bladder, there was also an increase in the effective force of gravity on the organism. That resulted in three receptor gene duplications – the parathyroid hormone-related protein receptor (PTHrPR), the glucocorticoid receptor (GR), and the ß-adrenergic receptor (ßAR). These receptor gene duplications were existential for the evolution of the lung, since all three resulted in amplification of lung physiologic signaling pathways necessary for air breathing. The PTHrPR was necessary for the formation of alveoli; ßAR amplification was essential for the pulmonary circulatory system to function independently from the systemic circulatory system, and the GR was vital for increasing the ßAR receptor density within the lung circulatory system. Selection pressure for PTHrPR expression in the water-land transition may have been initiated by its role in the calcification of bone, fortifying the skeleton to support the organism's increased weight on land. And selection pressure for the GR was due to increased blood pressure generated by gravitation on land causing shear stress on the microcirculation, offset by the evolution of the GR from the mineralocorticoid receptor, epistatically balancing the effect of the latter on blood pressure. The emergence of the GR would have facilitated the expression of ßARs, particularly in allowing for the independent regulation of the lung circulatory system, as mentioned above.

Epigenetic Inheritance

From this prima facie perspective, it is helpful to also consider the mechanism of epigenetic inheritance. It has been hypothesized that the phenotype is not just the form of the organism but is the actual means by which it expresses its "agency" to monitor the environment and detect changes that might prove to be existential threats. By recording such changes in the egg and sperm biochemically, modifying the DNA readout, the offspring are able to specifically adapt to such environmental changes using such foreknowledge (Nilsson et al. 2018).

This is the way the cell is able to "author" itself, the cell membrane delineating the outside of the cell from the inside. As the origin of the Mobius strip, the cell membrane "remembers" its history and is able to self-correct, referencing the lipids produced by pulsars in deep space, orienting themselves upward toward the sun and downward into the water, conducting thermal energy to fuel life.

References

D. Bohm, *Wholeness and the Implicate Order* (Routledge and Kegan Paul, London, 1980)

R.M. Case, D. Eisner, A. Gurney, O. Jones, S. Muallem, A. Verkhratsky, Evolution of calcium homeostasis: From birth of the first cell to an omnipresent signalling system. Cell Calcium **42**, 345–350 (2007)

D.E. Claassen, B.S. Spooner, Liposome formation in microgravity. Adv. Space Res. **17**, 151–160 (1996)

J.A. Clack, *Gaining ground* (Indiana University Press, Bloomington, 2012)

F. Demayo, P. Minoo, C.G. Plopper, L. Schuger, J. Shannon, J.S. Torday, Mesenchymal-epithelial interactions in lung development and repair: Are modeling and remodeling the same process? Am. J. Physiol. Lung Cell. Mol. Physiol. **283**, L510–L517 (2002)

S. Hawking, *A brief history of time* (Bantam, New York, 2011)

B. Kleine, W.G. Rossmanith, *Hormones and the endocrine system* (Springer, New York, 2016)

S. Korzh, C.L. Winata, W. Zheng, S. Yang, A. Yin, P. Ingham, V. Korzh, Z. Gong, The interaction of epithelial Ihha and mesenchymal Fgf10 in zebrafish esophageal and swimbladder development. Dev. Biol. **359**, 262–276 (2011)

C.J. Lang, A.D. Postle, S. Orgeig, F. Possmayer, W. Bernhard, A.K. Panda, K.D. Jürgens, W.K. Milsom, K. Nag, C.B. Daniels, Dipalmitoylphosphatidylcholine is not the major surfactant phospholipid species in all mammals. Am. J. Physiol. Regul. Integr. Comp. Physiol. **289**, R1426–R1439 (2005)

E.E. Nilsson, I. Sadler-Riggleman, M.K. Skinner, Environmentally induced epigenetic transgenerational inheritance of disease. Environ Epigenet **4**, dvy016 (2018)

S. Orgeig, W. Bernhard, S.C. Biswas, C.B. Daniels, S.B. Hall, S.K. Hetz, C.J. Lang, J.N. Maina, A.K. Panda, J. Perez-Gil, F. Possmayer, R.A. Veldhuizen, W. Yan, The anatomy, physics, and physiology of gas exchange surfaces: Is there a universal function for pulmonary surfactant in animal respiratory structures? Integr. Comp. Biol. **47**, 610–627 (2007)

C. Prem, W. Salvenmoser, J. Würtz, B. Pelster, Swim bladder gas gland cells produce surfactant: In vivo and in culture. Am. J. Physiol. Regul. Integr. Comp. Physiol. **279**, R2336–R2343 (2000)

B. Purevdorj-Gage, K.B. Sheehan, L.E. Hyman, Effects of low-shear modeled microgravity on cell function, gene expression, and phenotype in Saccharomyces cerevisiae. Appl. Environ. Microbiol. **72**, 4569–4575 (2006)

A.S. Romer, *The vertebrate story* (University of Chicago Press, Chicago, 1949)

N.H. Shubin, E.B. Daeschler, F.A. Jenkins Jr., The pectoral fin of Tiktaalik roseae and the origin of the tetrapod limb. Nature **440**, 764–771 (2006)

J. Slack, Role of fibroblast growth factors as inducing agents in early embryonic development. Mol. Reprod. Dev. **39**, 118–124 (1994)

S.J. Storr, C.M. Woolston, Y. Zhang, S.G. Martin, Redox environment, free radical, and oxidative DNA damage. Antioxid. Redox Signal. **18**, 2399–2408 (2013)

J.S. Torday, Pleiotropy, the physiologic basis for biologic fields. Prog. Biophys. Mol. Biol. **136**, 37–39 (2018)

Afterward

There is a prevailing sense that there is something greater than ourselves. I hope this book has helped to understand that that is because of our innate knowledge of the Cosmos, which is embedded in our physiology as cell-cell communications.

We have traversed the course of our evolutionary history backward in space and time to the unicell by exploiting the processes of cell-cell communication, mediated by soluble growth factors and their cognate receptors. Using that scientifically reductionist approach enables understanding of our origins as physical factors that have posed existential threats, coping with them by endogenization, compartmentalizing them, and making them functional as our physiology. Constructing such a biologic "network" that allows us to effectively interact with an ever-changing environment, embedding that history in our cellular-molecular makeup and retrieving that "memory" in order to rapidly and effectively solve future existential threats, is the power of the evolutionary process.

The book opens with subjective age, a phenomenon that gives insight to the role of the endocrine system in behaviors that span the life cycle. It closes with the recognition of the cell membrane as a Mobius strip, enabling the cell to "author" itself by remembering when inside and outside did not exist. The driver for life is negative entropy, which formed the basis for the ambiguity that impels us to interrogate our ever-changing environment. But without foreknowledge of our ontology and epistemology, we have made up "just-so stories" in place of the true story to fill in that void.

In contrast to that, the present perspective for "how and why" evolution has occurred teaches us that atoms and cells are homologues (of the same origin), both of which are deterministic and probabilistic, formed by the gravitational force originating from the Big Bang, constrained by homeostasis in order to comply with the laws of nature. It is in this way that we finally can realize that consciousness is not how matter sublimates as mind; it is the operating system common to both life and the cosmos.

J. S. Torday, *Hormones and Reality*, https://doi.org/10.1007/978-3-030-93691-4

Importantly, this novel way of understanding life from its beginning offers the insight that we are not primarily matter, we are energy, and that energy is literally the same as the energy that flows through everything in the Cosmos. Once we fully understand and appreciate our relationship to energy flows and consciousness, we will be able to realize the transition to more fully sentient human beings, motivated by cooperation with all our fellow beings, human and nonhuman alike, and our environment in order to remain synchronous with the vector formed by the Big Bang upon which our existence is based.

John S. Torday
Philadelphia, Pennsylvania
2021

Index

J. S. Torday, *Hormones and Reality*, https://doi.org/10.1007/978-3-030-93691-4